What is statistics?

Probability and statistics often go hand in hand. The way to consider the two is as follows:

Suppose there is a box containing 100 multi-colored ping pong balls. In a probability class, you would be told about the contents of the box:

50 red
30 white
20 blue

You would then be asked to determine the **probability** of pulling out a particular color combination if you randomly selected a certain number of balls from the box.

If you select 10 ping pong balls at random, what is the probability you will select 5 red, 3 white and 2 blue?

In a statistics class, you would not be told what about the contents of the box; rather, you would be told about the sample you have selected. You then would use that information to make inferences about what is in the box.

In a random sample of ten ping pong balls, four were blue. Find a 95% confidence interval for the proportion of all ping pong balls that are blue.

While the notion of a confidence interval may not make sense at this point, the overall meaning is that you would be asked to determine what proportion of the ping pong balls in the original box were blue. Obviously there is no way you would be able to determine exactly what proportion of the box consisted of blue ping pong balls (unless you counted them all,) but using **statistics** you can obtain a good estimate.

Preparation for the SOLs

Tree Diagram

Sample Space

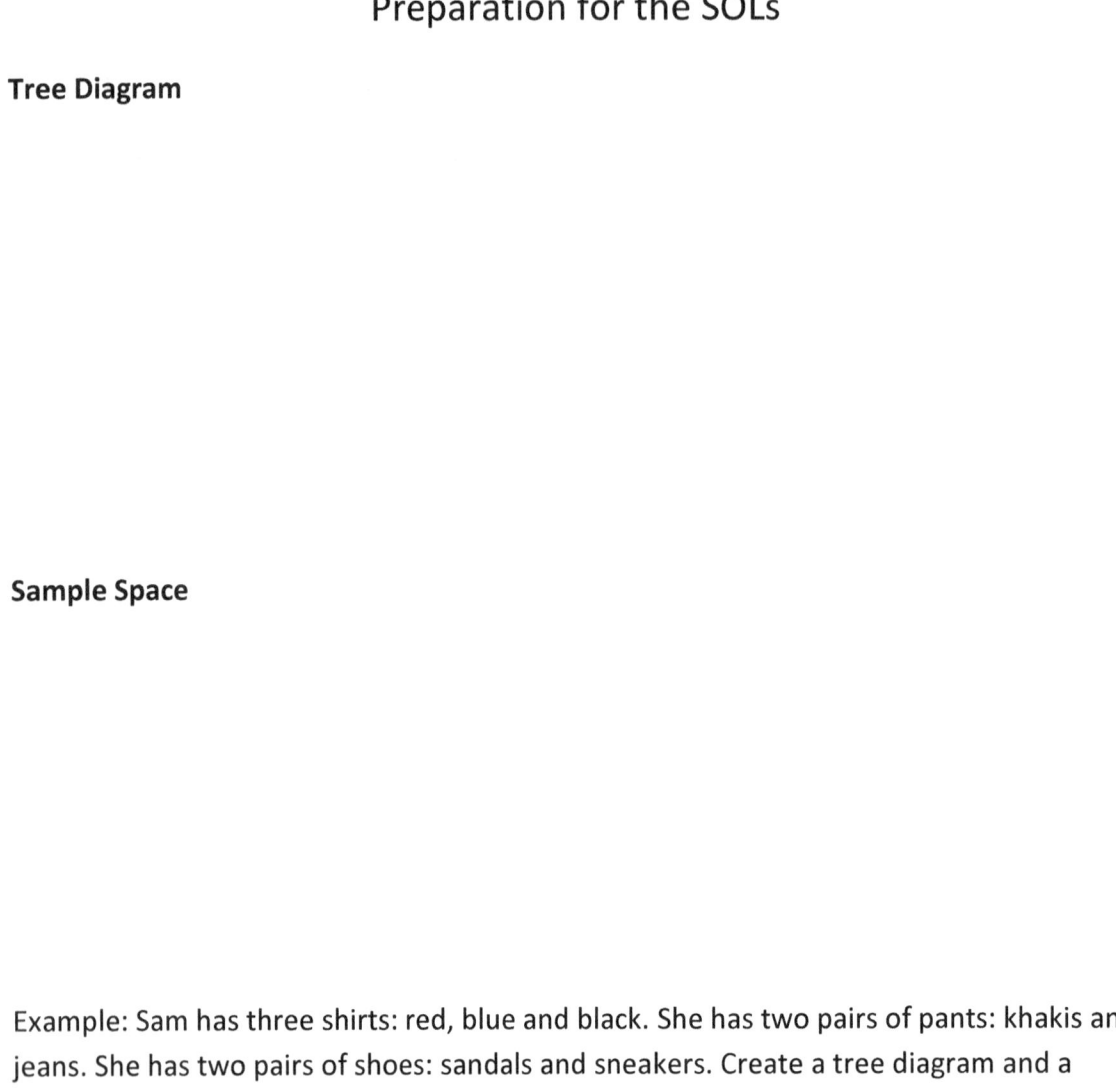

Example: Sam has three shirts: red, blue and black. She has two pairs of pants: khakis and jeans. She has two pairs of shoes: sandals and sneakers. Create a tree diagram and a sample space for the possible outfits she can make.

Example:

Ed has two different movies he'd like to see, a comedy and a drama. The times of each are 1:00, 3:00 and 7:00. Create a tree diagram and a sample space to demonstrate all of his possibilities.

Probability:

In the above example, what is the probability that Ed will see a drama if he randomly selects one of the choices? Express your answer as a fraction, decimal and a percent.

In the above example, what is the probability that Ed will see a comedy at 1:00 if he randomly chooses a movie? Express your answer as a fraction, decimal and a percent.

Fundamental counting principle:

If you have 7 shirt, 5 pairs of pants and 6 pairs of shoes, how many unique outfits can you make?

Basic Statistical Terms

Descriptive Statistics:

Inferential Statistics:

Constants:

Variables:

Qualitative:

Quantitative:

Continuous:

Discrete:

Determine if the following variables are qualitative, continuous quantitative or discrete quantitative:

The percent of students in this class who are juniors.

Your student ID number:

The number of students in the class:

The average number of students in all of your classes:

Your gender:

Vocabulary

Mutually Exclusive

Exhaustive

Independent

Mutually Exclusive, Independent and Exhaustive

Mutually exclusive:

Consider the following:

A class has the following make-up:

- A) Freshmen: 25%
- B) Sophomores: 30%
- C) Juniors: 35%
- D) Seniors:

1. If the class only consists of undergraduates, what percent are seniors?

2. If you select one student at random, what is the probability that the student is a sophomore or a senior?

3. If you select one student at random, what is the probability that the student is not a sophomore?

4. If you select one student at random, what is the probability that the student is a freshman and a senior?

Independent:

Consider the following for a class of undergraduates:

A) Freshmen: 25%   E) Blue eyes: 30%
B) Sophomores: 30%   F) Brown eyes: 50%
C) Juniors: 35%   G) Hazel: 15%
D) Senior: H) Other:

Fill in the missing amounts. What enables you to answer that question?

If you pick a student at random, what is the probability that the student will have both brown eyes and sophomore status?

If you pick a student at random, what is the probability that the student will have blue eyes and senior status?

If you pick two students at random, what is the probability that they will both be juniors (assume replacement)?

	Key word	Operation
Mutually exclusive		
Independent		

Examples:

A group of stat students were surveyed by being asked three different questions, and the partial results were as follows:

Question 1: Left-handed: 23% (A)

Question 2: Male: 52% (B)

Question 3: Current GPA ≥ 3.5: 34% (C)

$3.0 \leq$ Current GPA < 3.5 41% (D)

1. What percent of the class is right-handed?

2. If a student is selected at random, what is the probability the student will be male and have a GPA ≥ 3.5?

3. Find the probability of selecting a random student with a GPA < 3.0.

4. If you select two students at random, with replacement, what is the probability they will both have a GPA ≥ 3.5?

When is each graph appropriate?

Pie chart:

Line graph:

Histogram:

Bar graph:

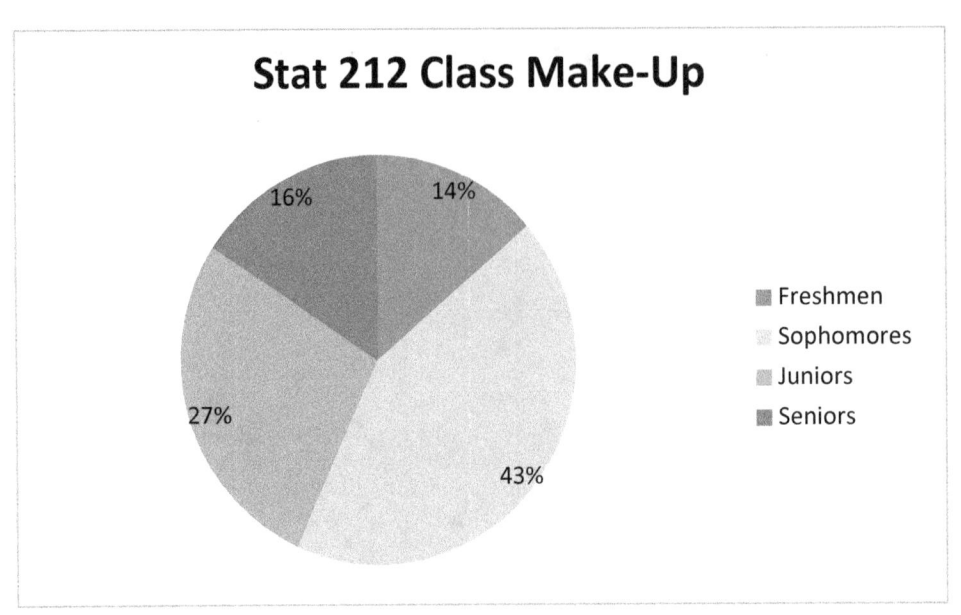

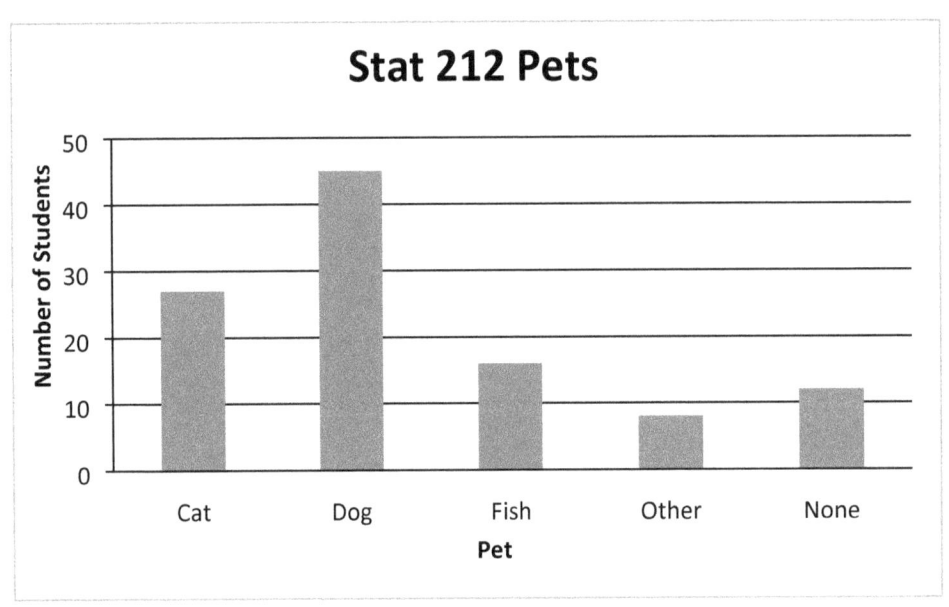

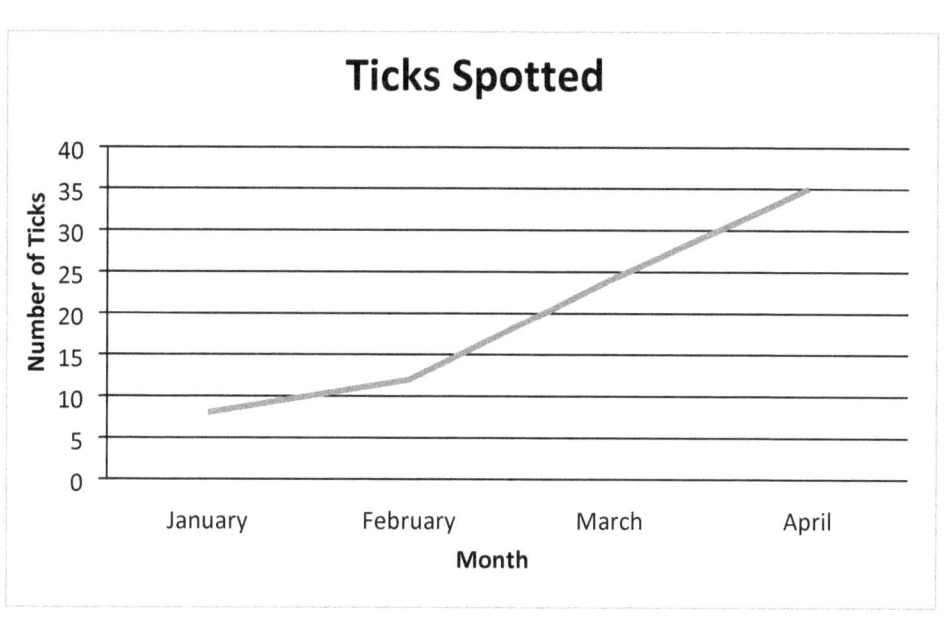

Which types of graphs are appropriate?

1. The puppy was weighed each on the first of each month for the first six months.

2. Students were asked which one core class was their favorite. All students were required to vote.

3. Students' heights were recorded.

4. Each third grade class was polled separately. The number of students who were only children was recorded for each class.

5. Students were asked what sports they play. The number of students who played each sport was recorded. Other and none were options.

6. Given the choice of four movies, students were able to pick which movie they liked the best. They were allowed to (and required to) vote only once.

Nominal, Ordinal and Interval

Nominal:

Ordinal:

Interval:

Determine if the following scenarios are nominal, ordinal or interval.

Enter 0 for no and 1 for yes:

On Goodreads.com, rate how well you liked the book on a scale from 1 to 5.

State which age range you fall into.

Rate how much pain you are in from a scale of 1 to 10.

Based on your race, classify yourself into category 1 through 7.

Determine what grade range you fall into.

Definitions

Population:

Individual:

Sampling frame:

Sample:

Sampling error:

Example:

Answer the questions that follow the scenario:

You wish to do a study to see how the children at your local elementary school feel about the quality of their education. You plan to go in to the school next Thursday and select 50 kids to be in your study. You will use that information to make statements about the whole school.

What is the **population** of interest?

Who are the **individuals** in the study?

What is the **sampling frame**?

Who makes up the **sample**?

Give an example of a potential **sampling error**.

Parameters vs. Statistics

Parameter:

Statistic:

In the following scenario, state whether the values given are parameters or statistics,

A local neighborhood consists of 576 homes. 385 of the households belong to the pool, resulting in a membership rate 66.84%. In order to see if people would be willing to pay for improvements to the pool, a random sample of 75 pool members was obtained. 48 people, or 64%, said that they would be willing to pay for some added amenities.

385

64%

75

48

Sampling Terms

Census:

Bias:

Selection Bias:

Non-response Bias:

Response Bias:

In the elementary school example, state what type of bias is present in the following scenarios:

The principal does the interviewing.

The gifted kids are on a field trip and cannot be selected for your sample.

You send a survey home with the fifty kids, and only thirty-seven respond.

The questions are worded poorly.

You select the students from the group of children who are involved in after-school activities.

Sample Collection Methods:

You wish to do a study to see how students at your local elementary school feel about the quality of their education.

If you were to do a **census**, what would you do?

For the other sample collection methods, assume you would like to obtain a sample of 50 students. Describe how you would obtain the sample using each method.

Simple Random:

Convenience:

Voluntary Response:

Systematic:

Cluster:	Stratified:	Multistage:

Sample collection activity

	Last Name	Gender	Score	Teacher	*
1	Bornheimer	M	483	Abernathy	
2	Button	F	495	Abernathy	
3	Camp	F	454	Abernathy	
4	Carroll	M	515	Abernathy	
5	Coleman	M	451	Abernathy	
6	Cordero	M	495	Abernathy	
7	D'Annuzio	F	481	Abernathy	
8	Fliegel	F	505	Abernathy	
9	Gunn	F	444	Abernathy	
10	Halpert	M	391	Abernathy	Remedial
11	Heidt	F	473	Abernathy	
12	Howard	M	486	Abernathy	
13	Kang	M	516	Abernathy	
14	McGee	M	531	Abernathy	
15	Overton	M	495	Abernathy	
16	Pao	F	487	Abernathy	
17	Petok	F	580	Abernathy	Gifted
18	Pohliv	M	514	Abernathy	
19	Poreba	F	510	Abernathy	
20	Sawyer	F	544	Abernathy	
21	Seargant	M	487	Abernathy	
22	Segal	F	507	Abernathy	
23	Sertich	F	447	Abernathy	
24	Siridivan	F	535	Abernathy	
25	Stodel	F	440	Abernathy	Remedial
26	Stupnitsky	M	572	Abernathy	Gifted
27	Vilaysack	M	463	Abernathy	ESL
28	Walsh	F	455	Abernathy	
29	Wang	M	530	Abernathy	
30	Weeks	M	503	Abernathy	
31	Woods	M	494	Abernathy	
32	Barnes	F	460	Bauer	
33	Brett	F	508	Bauer	
34	Buckley	M	580	Bauer	Gifted
35	Carlson	M	498	Bauer	
36	Chowdhry	M	445	Bauer	ESL

37	Cordero	M	**580**	Bauer	Gifted
38	Elba	M	**522**	Bauer	
39	Esteves	M	**503**	Bauer	
40	Gordon	F	**521**	Bauer	
41	Jones	F	**553**	Bauer	
42	Koechner	M	**523**	Bauer	
43	Lashaway	M	**525**	Bauer	
44	Lindsay	M	**458**	Bauer	
45	Mackey	F	**549**	Bauer	
46	Nelson	F	**510**	Bauer	
47	Ogasaki	F	**492**	Bauer	
48	Park	M	**486**	Bauer	
49	Phelps	M	**478**	Bauer	
50	Pine	M	**424**	Bauer	Remedial
51	Proksch	M	**480**	Bauer	ESL
52	Purl	F	**512**	Bauer	
53	Robbins	M	**516**	Bauer	
54	Rosenberg	F	**497**	Bauer	
55	Schafer	M	**468**	Bauer	
56	Schur	M	**433**	Bauer	Remedial
57	Shafer	M	**481**	Bauer	
58	Shouldis	M	**437**	Bauer	Remedial
59	Stewart	F	**580**	Bauer	Gifted
60	Turner	M	**506**	Bauer	
61	Walls	F	**479**	Bauer	
62	Watson	F	**477**	Bauer	
63	York	M	**472**	Bauer	
64	Broad	F	**513**	Ellsworth	
65	Baker	M	**580**	Ellsworth	Gifted
66	Barton	M	**550**	Ellsworth	
67	Baumgartner	M	**487**	Ellsworth	
68	Carter	M	**506**	Ellsworth	
69	Denman	M	**506**	Ellsworth	
70	Fisher	F	**489**	Ellsworth	
71	Flannery	F	**507**	Ellsworth	
72	Hardin	F	**426**	Ellsworth	Remedial
73	Helms	M	**495**	Ellsworth	
74	Kaling	F	**519**	Ellsworth	
75	Kemper	F	**564**	Ellsworth	Gifted
76	Kinsey	F	**580**	Ellsworth	Gifted
77	Leiberstein	M	**510**	Ellsworth	

78	Novak	M	**493**	Ellsworth	
79	Nunez	M	**565**	Ellsworth	Gifted
80	Orlowski	M	**520**	Ellsworth	
81	Robinson	M	**537**	Ellsworth	
82	Ryan	F	**408**	Ellsworth	Remedial
83	Smith	F	**553**	Ellsworth	
84	Speicher	M	**467**	Ellsworth	
85	Talbot	F	**519**	Ellsworth	
86	Tan	M	**558**	Ellsworth	Gifted
87	Taylor	M	**395**	Ellsworth	Remedial
88	Townsend	F	**474**	Ellsworth	
89	Woods	M	**527**	Ellsworth	

If you wish to take a stratified random sample of size 30, determine how many students of each description you would need:

Gifted:

Remedial:

ESL:

Traditional:

Sample collection methods

If you were to use a phone book to collect a sample, label the method you have chosen for the following descriptions:

1. You select 10 pages at random, and use 10 randomly selected names from each of those 10 pages.

2. You randomly select three pages and use every name from each of those three pages.

3. You use every name in the book.

4. You use the first 100 names listed.

5. You two randomly selected names from each page.

6. You associate each name with a number and randomly generate 200 numbers; you use the names associated with those numbers.

7. You use the top name from every tenth page.

Design of Experiments

Experimental units:

Factor:

Levels:

Factor and levels example: A teacher wants to optimize student performance. There are three curriculums to choose from, and the option to take the course in-person and online are both available. How many factors are there? How many levels?

Treatment group:

Placebo:

Control group:

Replication:

Confounding variables:

Randomization:

Double blind:

Single blind:

Completely randomized design:

Block design:

Matched pairs design:

Design examples:

What types of designs are used in the following scenarios?

An advertising agency wants to see the effectiveness of a new ad it has created, which was designed to persuade people to vote a particular way. Three-hundred registered voters were randomly selected, and one-hundred-fifty of those people were randomly chosen to watch the ad. Each group was then polled to determine what percent of the sample will be voting the way the ad had encouraged.

Forty students were signed up for a tennis clinic with two instructors. (The clinic ran from 3 to 5 on Tuesdays and Thursdays.) The owner of the clinic was interested in determining if one instructor's methods were more effective than the other's. Twenty students were randomly assigned to each of the instructors. At the end of the program, the students were tasked with playing against each other: Instructor A's top student played against Instructor B's top student, the second ranking students played each other, and so on down through number twenty. The number of wins was recorded for each instructor.

A new drug has been created to ease the symptoms of diabetes. In order to test the drug, a randomly selected sample of 500 diabetics was divided into two groups based on whether they had type one or type two diabetes. Within each group, half of the subjects were randomly chosen to take the drug, while the rest were given the placebo. At the end of the study, the improvements were noted within each group.

A teacher wishes to test the effectiveness of a program which claims to improve SAT scores by an average of 100 points. A group of 100 randomly selected students who had already taken the SATs was collected, and those students went through the program. The students took the SATs again, and the average improvement was calculated.

Design of experiments example:

In question is the effectiveness of a new weight loss drug when compared to a current drug called *Body Slim*. 160 people were randomly selected to receive one of several treatment paths.

One quarter of the participants were told to take *Body Slim*, 1/4 were told to take the experimental drug, 1/4 were given a placebo, and 1/4 had nothing.

Within each group, half of the participants were told to exercise 5 hours per week; others were told to exercise ten.

Within each group, half of the participants were told to limit their diet to 1200 calories; the rest were told to consume 2000.

In the end, the amount of weight lost was computed for each person, and each group's average was compared to the others.

Who are the experimental units?

What are the factors?

How many levels are there?

What are the treatment groups?

What are the control groups?

Is this single or double blind?

Is this completely randomized, block, or matched pairs design?

Data Shapes

Bell curve:

Skewed to the right:

Skewed to the left:

Unimodal:

Bimodal:

Shape	Center	Spread

Stem and Leaf plots

Create a stem and leaf plot for the times for runners to complete a race, in minutes.

49	55	38	65	58	67	49	43	38	71	43	48
	29	56	58	51	63	47	52				

What shape is the data?

Create a stem and leaf plot for the following salaries from a single department at a local company.

41,612	40,901	41,215	41,613	40,389	42,284
41,658	40,716	43,289	42,517	41,381	42,049
40,611	44,921	42,613	41,391	41,722	41,300

What is the shape of the data?

A farmer wants to compare two fertilizers, A and B, to see if there is a difference in the amount of potatoes produced. He divides each of the same-sized plots exactly in half, using both fertilizers on each plot. The weights of each yield are noted and listed below in pounds, listed in no particular order.

Fertilizer A:

141.9	166.3	131.4	181.1	123.2	151.9
157.5	148.5	138.9	152.2	126.1	117.2
130.2	175.2	164.1	146.1	144.2	

Fertilizer B:

171.4	172.8	141.4	191.4	156.5	181.7
159.9	161.5	171.9	186.5	201.6	183.8
162.1	177.1	192.9	164.3	177.3	

Create a two-sided stem and leaf plot to compare the yields.

Does there appear to be a difference between the fertilizers?

Measures of Central Tendency

Write the variables that are used to describe each value, and list whether it is a parameter or a statistic.

Population mean:

Sample mean:

Population median:

Sample median:

Population standard deviation:

Sample standard deviation:

Population variance:

Sample variance:

Rule of thumb: Greek letters describe the population and are parameters. English letters describe the sample and are statistics.

Find the median, 5# summary, range and IQR

Stem	Leaves
1	4
2	
3	
4	6
5	46
6	0273
7	4418
8	2

1. Put the numbers in order

14   46   54   56   60   62   63   67   71   74   74   78   82

2. Find the median

a) $\dfrac{n+1}{2} = \dfrac{13+1}{2} = \dfrac{14}{2} = 7$ 7th slot is the median

14   46   54   56   60   62   **63**   67   71   74   74   78   82

3. If the median is an actual value, cross it out. If the median is "between" two numbers, draw a vertical line between them to separate the halves. Then average the two numbers together to get the value of the median.

14   46   54   56   60   62   **6|3**   67   71   74   74   78   82

4. Look at only the LEFT half of the data set and repeat steps 1 and 2 to find the first quartile.

$$\frac{n+1}{2} = \frac{6+1}{2} = 3.5 \qquad\qquad 3.5^{\text{th}} \text{ slot is the first quartile}$$

14   46   54 | 56   60   62

Since the first quartile is between two values, average those values together to get the value of Q1

$$\frac{54 + 56}{2} = 55$$

5. Repeat the same process on the right side to get the third quartile

$$\frac{n+1}{2} = \frac{6+1}{2} = 3.5 \qquad\qquad 3.5^{\text{th}} \text{ slot is the third quartile}$$

67   71   74 | 74   78   82

Since the first quartile is between two values, average those values together to get the value of Q3

$$\frac{74 + 74}{2} = 74$$

Five number summary:

Min: 14

Q1:   55

Med: 63

Q3:   74

Max: 82

Range = Max – Min = 82 – 14 = 68

IQR = Q3 – Q1 = 74 – 55 = 19

Median and five number summary example

A 25 year old woman recently went on match.com, and she recorded the ages of the men who contacted her, in the order they were received. The ages were as follows:

25	27	28	24	23	33	30	20	72	23	28	25
25	26	31	27	26	28	29	26				

First create a stem and leaf plot to determine the shape of the data.

Since this data is skewed (has an outlier,) we want to find the median as the measure of center.

First step: Rank the data from smallest to largest.

L(M) = the location of the median, and it is found by:

$$L(M) = \frac{n+1}{2}$$

Where n is the sample size. Find the location of this median.

Find the value of the median.

Suppose another value gets added to this data set:

20 23 23 24 25 25 25 26 26 26 27 27 28 28 28 29 30 31 33 **40** 72

Now find the L(M) and the value of the median.

Using the original data set, find the five number summary:

20 23 23 24 25 25 25 26 26 26 27 27 28 28 28 29 30 31 33 72

Range:

IQR:

Create a box and whisker plot:

Fences:

Lower fence formula:

Upper fence formula:

Adjacent values:

10   20   30   40   50   60   70

2. The following are unemployment rates for 12 randomly selected states.

Alabama	6.4
Idaho	4.2
Michigan	12.5
Connecticut	8.7
California	7.5
New Mexico	6.8
New Jersey	5.0
Ohio	8.2
Texas	7.2
North Dakota	4.1
New Hampshire	5.2
Mississippi	6.7

Create a stem and leaf plot; determine the five number summary, range and IQR.

Calculate the fences and draw a box and whisker plot.

4	5	6	7	8	9	10	11	12

Mean

How to find mean:

How to find mean on the TI83/84

Enter data into one of the lists.

Press the [STAT] button again and arrow over to CALC.

Select option 1 and hit [ENTER].

Select which list (L_1 or L_2 for instance) you would like analyzed. L_1 is the default if left blank.

Hit [ENTER]

You will be given the mean, standard deviation and five number summary

Example:

The following are the times, in minutes, for a random sample of runners to complete a race. You have made the stem and leaf plot before and determined that these times form a bell shaped distribution.

49	55	38	65	58	67	49	43	38	71	43	48
	29	56	58	51	63	47	52				

Find the mean time for this sample of runners.

What variable would be used to represent this value?

Is this a parameter or a statistic?

Can we say the average time for all runners is exactly this value? Why or why not?

Trimmed mean

Determine the mean of this data set (from dating website example.)

20 23 23 24 25 25 25 26 26 26 27 27 28 28 28 29 30 31 33 72

Does that value represent the center of the distribution?

Trimmed mean (definition):

Removing the two extreme values, one from each end, what is the new mean?

Trimming percentage:

$$\frac{\# \text{ of values removed from each end}}{n} \text{ x } 100$$

Determine the trimming percentage in this example.

How many values would have to be removed from each end to create a 10% trimming percentage? What would the new mean be?

Standard Deviation:

The standard deviation represents a typical difference between a data value and the mean.

Example: The IQs of 12 randomly selected people were recorded and are as follows:

101   104   96   109   89   79   99   117   94   111   106   95

Calculate the mean of this sample:

What variable is used to represent this value?

Is this a statistic or a parameter?

Fill out the following table:

x_i	$x_i - \overline{X}$	$(x_i - \overline{X})^2$

Sum:

What is the sum of the squared deviations?

Find the degrees of freedom:

Divide the sum of the squared deviations by the degrees of freedom:

That value is the *variance*.

Take the square root of that value. That is the *standard deviation*.

Variance formula = $\dfrac{\sum (x_i - \bar{X})^2}{n - 1}$

Standard deviation formula = $\sqrt{\dfrac{\sum (x_i - \bar{X})^2}{n - 1}}$

Finding standard deviation on the TI 83/84 is done the same way as the mean.

Drawing and labeling normal (bell) curves

IQs are normally distributed in the population. Use the findings from your sample to draw and label the normal curve:

Mean =

Standard deviation =

How many of the values fell within one standard deviation of the mean? What percent was that?

How many values fell within two standard deviations of the mean? What percent is that?

How many of the values fell within three standard deviations of the mean? What percent is that?

Empirical Rule

68-95-99.7 Rule

68% of the data falls within one standard deviation of the mean.

95% of the data falls within two standard deviations of the mean.

99.7% of the data falls within three standard deviations of the mean.

Mean/standard deviation examples

A local elementary school timed how long it took students to sprint from one end of the playground to the other. A random sample of 18 runners was selected, and their times (in seconds) are as follows:

34	59	19	25	45	42	54	31	35	27	38	41
	60	46	45	48	39	51					

Calculate the mean and the median of the data set:

Based on those values, what shape is the data?

Calculate the 5% trimmed mean. How does the trimmed mean compare to the mean? Why was that the case?

Draw and label this distribution for test scores: The shape is bell, $\mu = 76.3$, and $\sigma = 7.1$

95% of students scored between?

Scatterplots

Explanatory variable:

Response variable:

Positive correlation:

Negative correlation:

Determine if the following correlations are positive, negative or zero.

The age of a child and the grade the child is in.

The number of hours spent studying and the grade on the test.

Adult height and college GPA.

The age of your car and the value of your car.

The number of students in a class and the performance of the students on standardized tests.

The number of people at the mall and the number of available parking spaces.

The number of teachers at a school and the number of injuries on the playground.

How to graph a scatterplot on the TI 83/84:

Once the data has been entered in two of the lists, hit [2ND] then [y=] to select the STATPLOT option.
Turn the Statplot ON.
Select the LINE GRAPH or SCATTERPLOT option from the choices of TYPE.
Make sure you select the correct list (L_1 or L_2 for instance) for Xlist.
Make sure you select the correct list (L_1 or L_2 for instance) for Ylist
Choose which mark you would like.
Adjust the WINDOW so that the graph will be visible on the screen.
Press [GRAPH]

Consider the following information regarding the horsepower of car engines versus the miles per gallon.

Horsepower	MPG
410	14
370	15
350	17
322	19
309	20
270	22
253	24
225	29

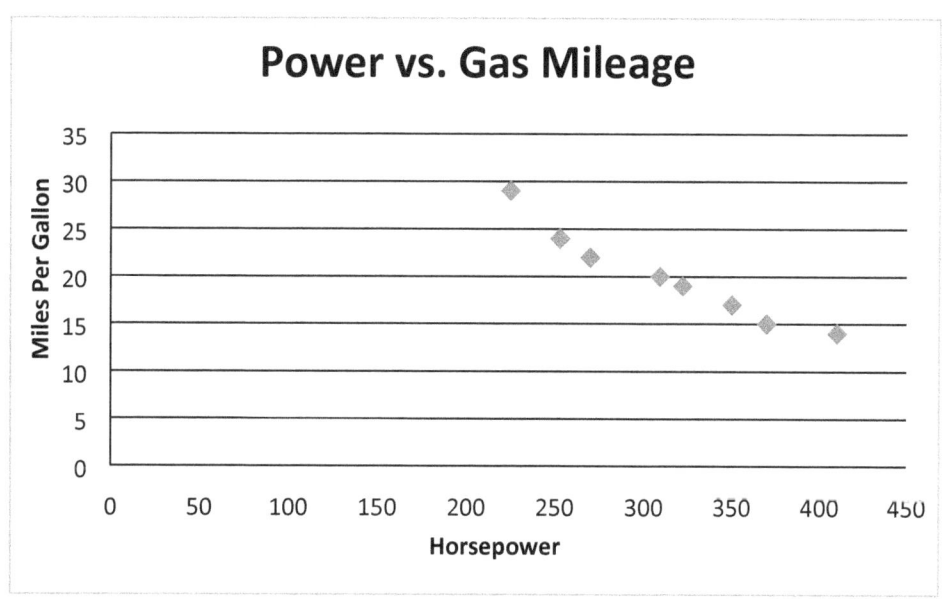

Is the correlation positive or negative? Explain.

Is the relationship strong or weak? Explain.

Estimate what the value of the correlation coefficient will be.

Calculating Correlation Coefficient

Horsepower	MPG			
x_i	y_i	x_i^2	y_i^2	$x_i y_i$
410	14	168100	196	5740
370	15	136900	225	5550
350	17	122500	289	5950
322	19	103684	361	6118
309	20	95481	400	6180
270	22	72900	484	5940
253	24	64009	576	6072
225	29	50625	841	6525
$\sum x_i$	$\sum y_i$	$\sum (x_i^2)$	$\sum (y_i^2)$	$\sum (x_i y_i)$
2509	160	814199	3372	48075

$$ r = \frac{S_{xy}}{\sqrt{S_{xx} S_{yy}}} $$

$$ S_{xx} = \sum (x_i^2) - \frac{(\sum x_i)^2}{n} \quad = 814199 - \frac{(2509)^2}{8} \quad = 27313.875 $$

$$ S_{yy} = \sum (y_i^2) - \frac{(\sum y_i)^2}{n} \quad = 3372 - \frac{(160)^2}{8} \quad = 172 $$

$$S_{xy} = \sum(x_i y_i) - \frac{(\sum x_i)(\sum y_i)}{n} = 48075 - \frac{(2509)(160)}{8} = -2105$$

$$r = \frac{-2105}{\sqrt{(27313.875)(172)}} = -0.97117$$

Calculating Correlation Coefficient on TI 83/84

How to determine correlation coefficient on the calculator:

Enter the data in two of the lists.

Select [2ND] and [0] to access the CATALOG.

Select DiagnosticOn and hit [ENTER] twice. (The word DONE should appear).

Press the [STAT] button again and arrow over to CALC.

Select option 8 and hit [ENTER].

Select which lists (L_1 or L_2 for instance) you would like analyzed, separated by a comma.

The first list entered is the X (independent) variable. $L_1 = x$ and $L_2 = y$ are the defaults if left blank.

Hit [ENTER].

You will see r, r^2, and the slope and the intercept of the regression equation.

Use the previous example to calculate r on the calculator:

Determine the correlation coefficient for the following scatterplot. Graph it on the calculator, and describe the plot with respect to strength, direction and linearity.

Age:	Weight:	Age:	Weight:	Age:	Weight:
30	250	30	127	25	182
25	200	40	235	40	108
42	162	45	191	30	196
32	167	37	202	41	230

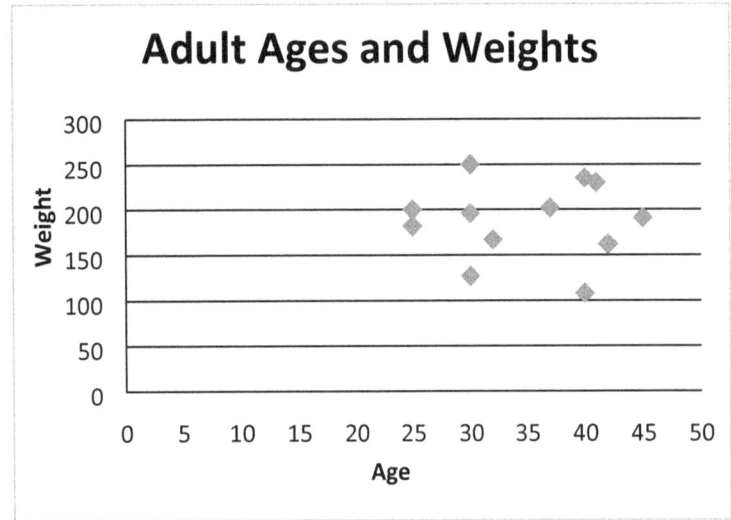

Determine the correlation coefficient for the following scatterplot. Graph it on the calculator and describe the plot with respect to strength, direction and linearity.

Hours	Growth
0	3.1
1	3.8
2	4.4
3	5.1
4	5.9
5	6.1
6	6
7	5.3
8	4.5
9	3.9
10	3.2

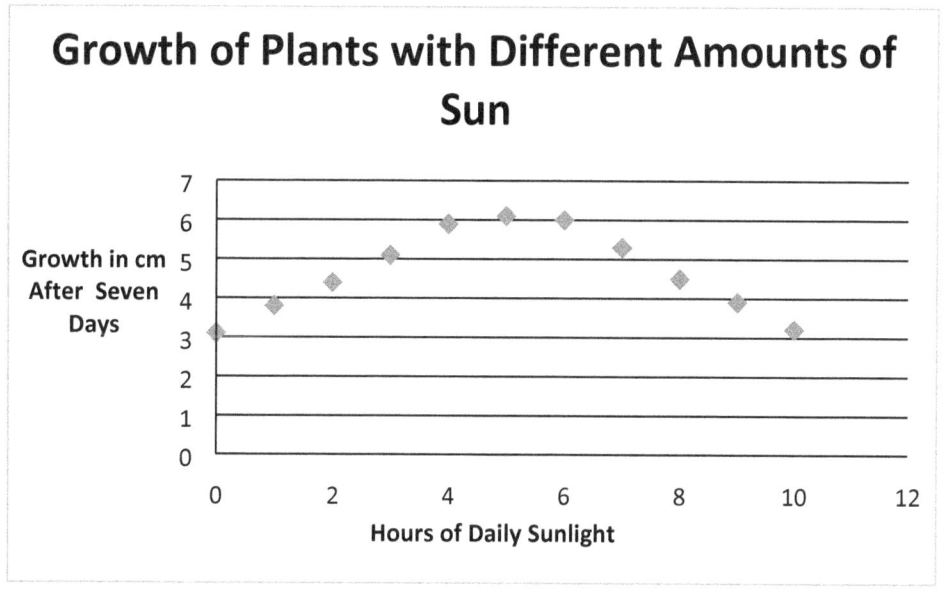

What type of equation would best fit these points?

Least Squares Regression

In order to use scatterplots to predict average y values when given an x, we need to find the equation of the "line of best fit." This line minimizes the vertical distance between each point and the line.

Example:

Consider the horsepower/gas mileage example.

Horsepower	MPG
410	14
370	15
350	17
322	19
309	20
270	22
253	24
225	29

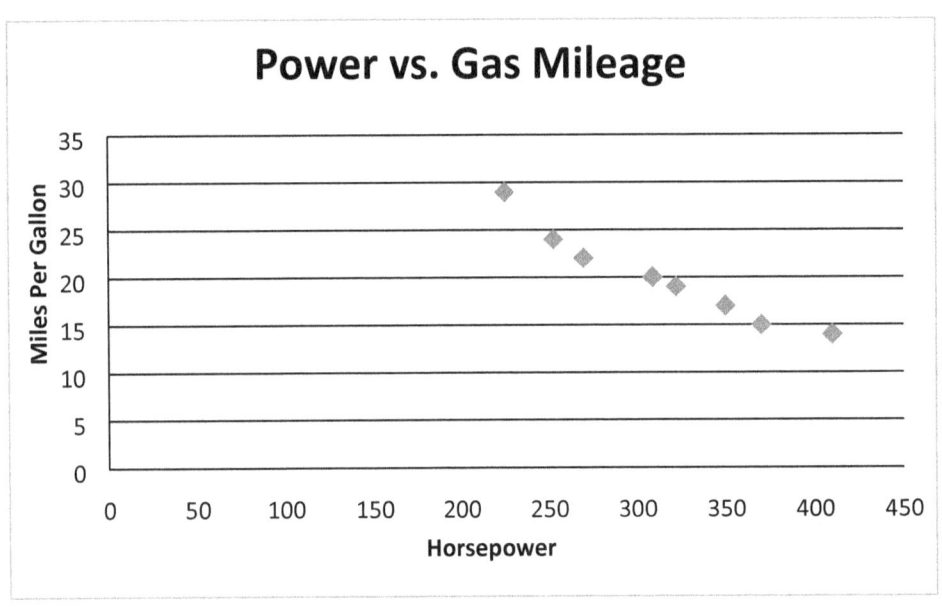

Zooming in on the graph, we get:

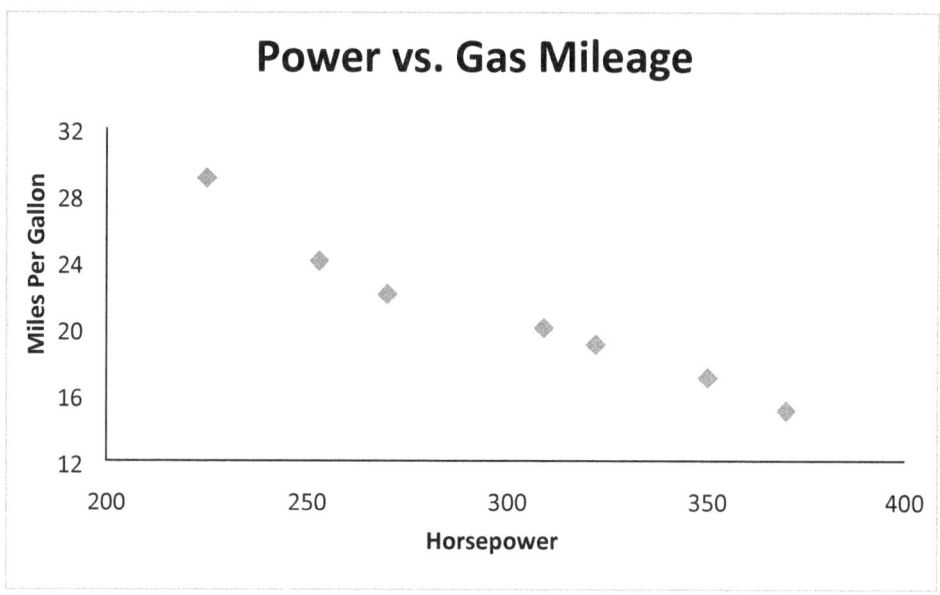

We want to find the line that minimizes the vertical distances between the points and the line.

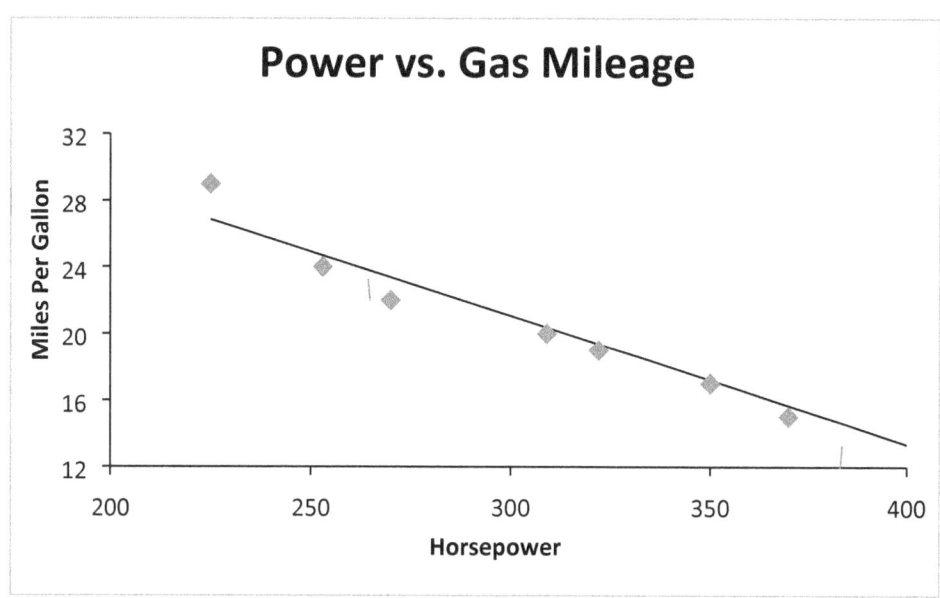

Each vertical distance is known as a residual and is represented as follows:

$$y - \hat{y}$$

Where y is the observed value and $\hat{y}$ is the predicted value (from the line).

Values below the line have a negative residual; values above the line have a positive residual. The sum of the residuals is always 0.

In order to make sure the residual values are always positive, the residuals get squared. We are looking for the line that will minimize the sum of the squared residuals.

In other words, we look to minimize $\sum (y - \hat{y})^2$

The equation that will minimize $\sum (y - \hat{y})^2$ is:

$$y = \beta_0 + \beta_1 x$$

Where β_0 is the intercept and β_1 is the slope.

$$\beta_1 = \frac{S_{xy}}{S_{xx}}$$

And

$$\beta_0 = \bar{y} - \beta_1 \bar{x}$$

Example:

Consider the horsepower/gas mileage example.

Horsepower	MPG
410	14
370	15
350	17
322	19
309	20
270	22
253	24
225	29

Remember:

$S_{xx} = 27313.875$ and $S_{xy} = -2105$

Therefore:

$$\hat{\beta}_1 = \frac{-2105}{27313.875} \approx -.0770670584$$

To determine $\hat{\beta}_0$ we would need to know $\bar{x}$ and $\bar{y}$, which are 313.625 and 20, respectively.

$$\hat{\beta}_0 = \bar{y} - \hat{\beta}_1 \bar{x} = 20 - (-.0770670584)(313.625) \approx 20 + 24.1702 = 44.1702$$

So the equation for the least squares regression line is:

$$y = 44.1702 - .0771x$$

How to find that regression line in the TI 83/84:

Enter the data in two of the lists.

Select [2ND] and [0] to access the CATALOG.

Select DiagnosticOn and hit [ENTER] twice. (The word DONE should appear).

Press the [STAT] button again and arrow over to CALC.

Select option 8 and hit [ENTER].

Select which lists (L_1 or L_2 for instance) you would like analyzed, separated by a comma.

The first list entered is the X (independent) variable. $L_1 = x$ and $L_2 = y$ are the defaults if left blank.

Hit [ENTER].

You will see r, r^2, and the slope and the intercept of the regression equation.

How to graph the regression line and the scatter plot together:

Press [Y=]

Type in the equation given above, using [X,T,θ,n] for X

Make sure the STAT PLOT is turned on

Press [GRAPH]

How to use the regression line to predict y values given an x:

Press [2nd][WINDOW] for TBLSET

Determine which X value you would like to start with (TblStart)

ΔTbl= The increment you would like to increase by

Indpnt: Auto

OR

Turn Indpnt to Ask to choose certain x values.

Select [2nd] [GRAPH] for TABLE

Input the desired x values

Example:

The following values reflect the average monthly high temperature and the amount of the gas bill (heating bill).

Month	Avg. High	Gas Bill
January	32	189
February	29	204
March	36	172
April	48	150
May	62	98
June	74	60
July	82	34
August	85	36
September	73	67
October	55	105
November	46	142
December	39	199

Before graphing, specify whether the correlation will be positive or negative, and explain why.

Graph the scatterplot on the calculator. Describe the correlation.

What is the correlation coefficient?

What is the equation of the regression line?

What would be the predicted bill of a month with an average high of 50 degrees?

What would be the predicted bill of a month with an average high of 96 degrees? Why is that answer unreliable?

Complete the sentence: For every degree increase in average temperature…

Outliers, Influential Observations and Coefficient of Determination

Coefficient of determination:

Outlier:

Influential observation:

Consider the following table which relates the average number of hours of exercise per week and the total weight loss (in pounds) after four weeks:

Average number of hours of exercise per week	Total weight loss (lbs) after four weeks
4.1	5.2
3.2	6.7
6.1	8.1
8.4	9.2
10.9	10.4
5.4	8.7
6.7	6.2

The graph looks like this:

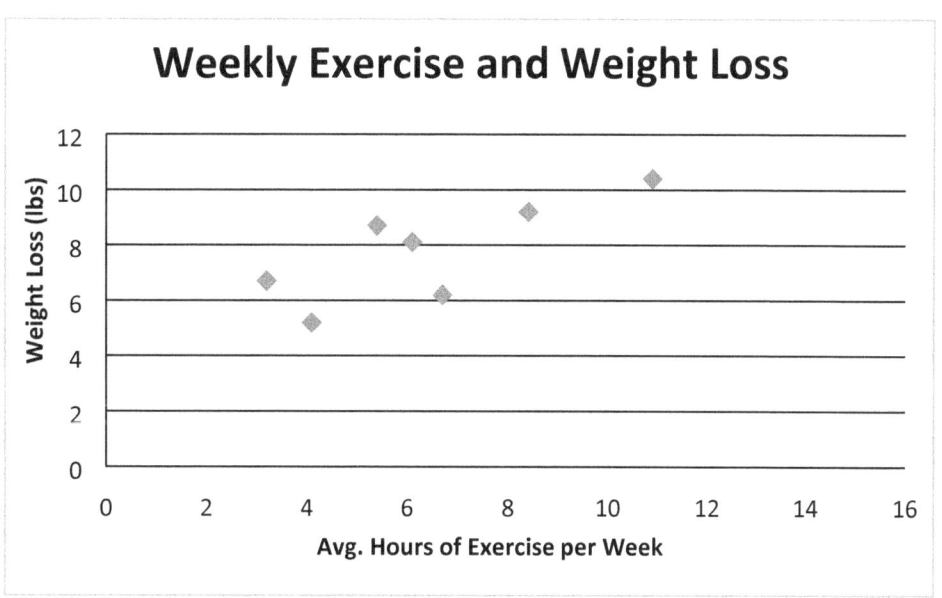

What is the correlation coefficient?

What is the coefficient of determination?

What is the formula for the regression line?

With the regression line plotted, the graph looks like this:

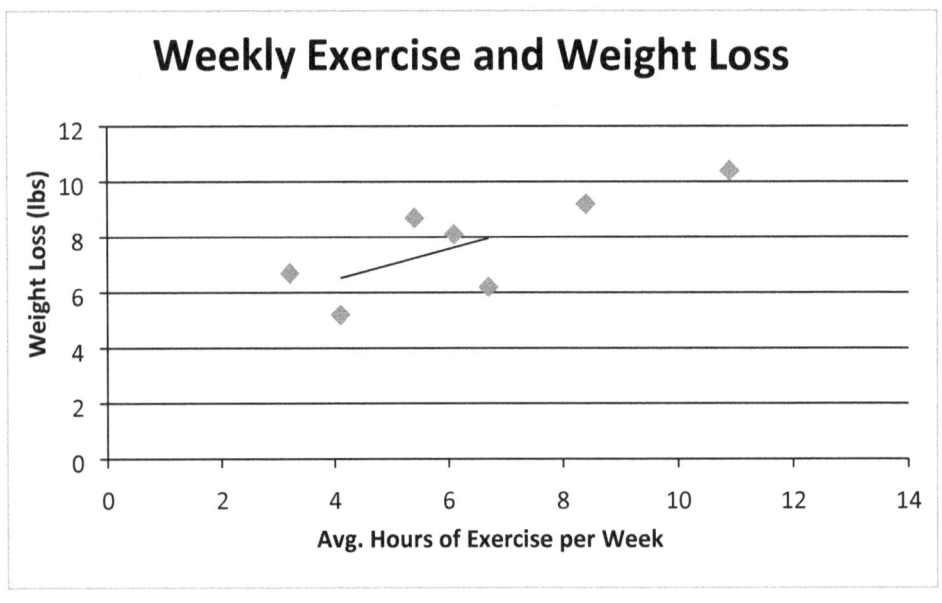

Now add one more value to the table:

Average number of hours of exercise per week	Total weight loss (lbs) after four weeks
4.1	5.2
3.2	6.7
6.1	8.1
8.4	9.2
10.9	10.4
5.4	8.7
6.7	6.2
7.6	**2**

The new graph looks like this:

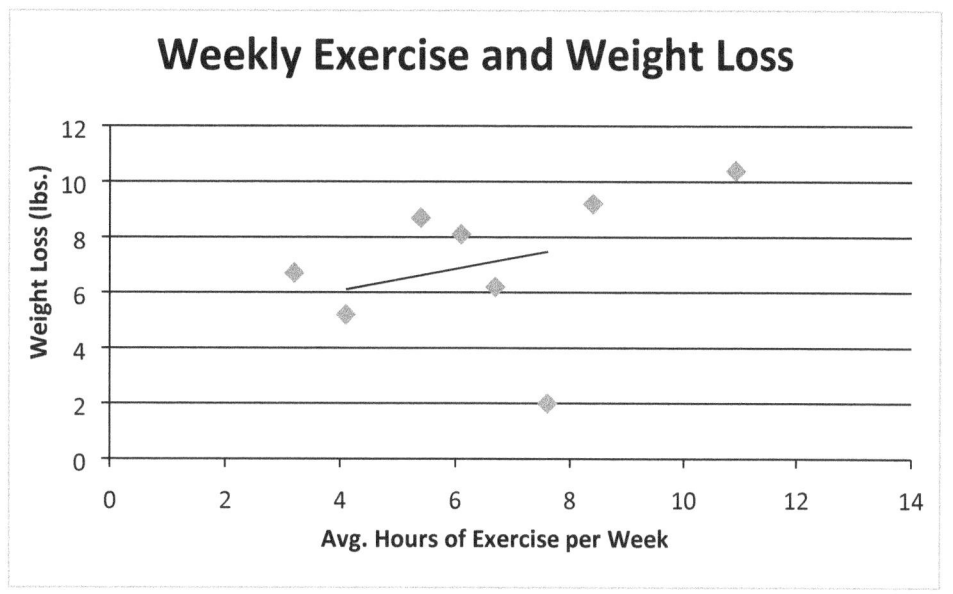

What is the correlation coefficient?

What is the coefficient of determination?

What is the equation of the regression line?

What does the slope of that regression line tell you?

Is the point (7.6, 2) an influential observation or an outlier?

Example 2:

Consider this table that relates the square footage of houses to their cost:

Square Footage	Price in Thousands
1420	174
1567	129
1624	158
1759	136
1838	149
1964	168

The graph looks like this:

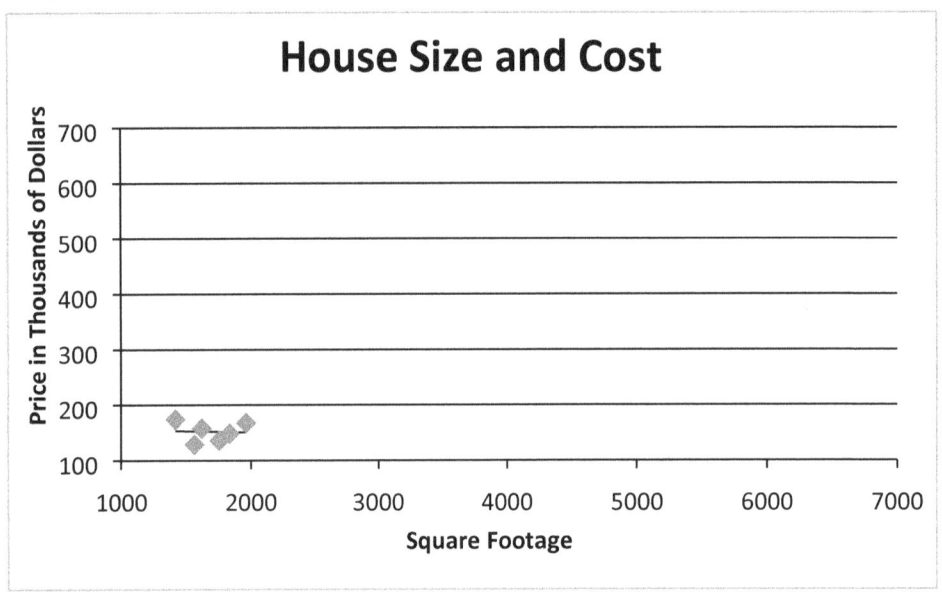

What is the correlation coefficient?

What is the coefficient of determination?

What is the equation of the regression line? What does the slope of that line mean?

Does it appear that a strong relationship exists between the size of the house and its value?

Now consider the same data with one additional value:

Square Footage	Price in Thousands
1420	174
1567	129
1624	158
1759	136
1838	149
6429	**650**
1964	168

The graph now looks like this:

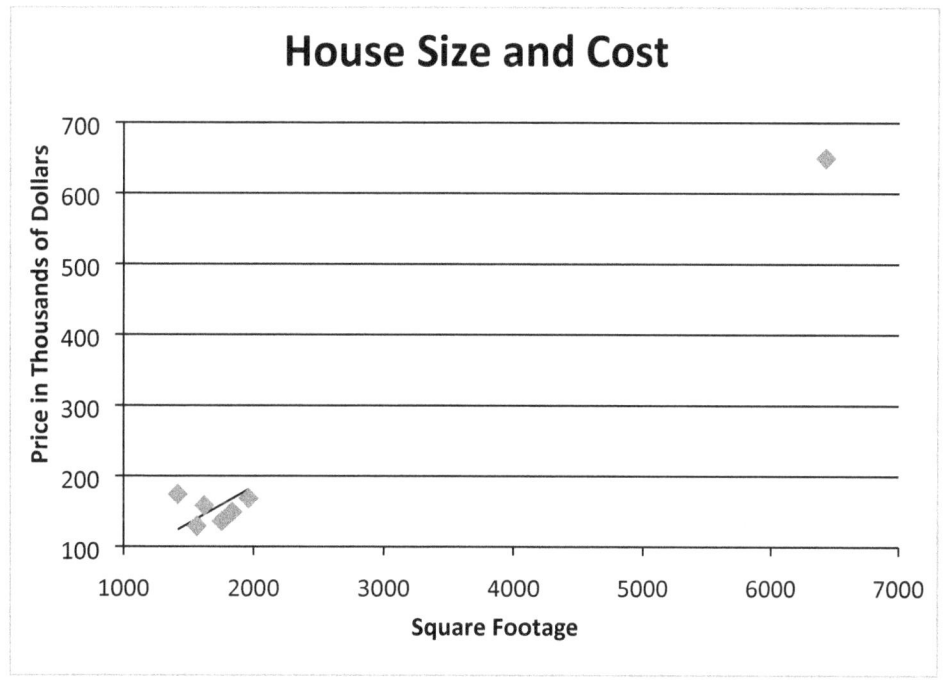

What is the correlation coefficient?

What is the coefficient of determination?

What is the equation of the regression line? What does the slope of that line tell you?

Is this an influential observation or an outlier?

Z-score transformations

Goals:

1. Know 68 – 95 – 99.7 rule

2. Convert actual values into percentiles

3. Convert percentiles into actual values

Actual ⟶ Z ⟶ Percentile

Percentile ⟶ Z ⟶ Actual

Drawing and labeling normal curves
68-95-99.7 rule

The students in **CLASS A** took an exam, and the mean was 70 with a standard deviation of 10.
Draw and label that normal curve above the appropriate spot on the number line:

CLASS B had a mean of 70 and a standard deviation of 2.
Draw and label that normal curve above the appropriate spot on the number line:

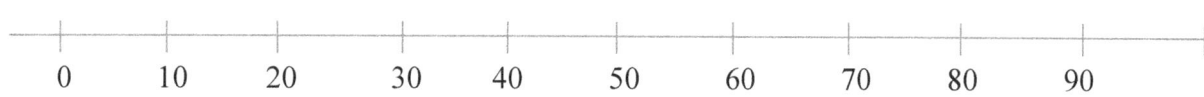

CLASS C had a mean of 50 and a standard deviation of 10.
Draw and label that normal curve above the appropriate spot on the number line:

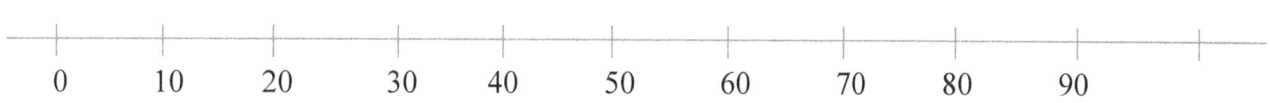

What impact did decreasing the standard deviation have? (Class A to B)

What impact did changing the mean have? (Class A to C)

The 68-95-99.7 rule:

1. 68% of the data falls within one standard deviation of the mean in each direction.

2. 95% of the data falls within two standard deviations of the mean in each direction.

3. 99.7% of the data falls within three standard deviations of the mean in each direction.

Refer to the previous page to answer the following questions:

1. What percent of the students in class A scored between a 50 and a 90?

2. 68% of the students in class C scored between which two values?

3. 99.7% of the students from class B scored between which two values?

Z-score Transformations:

Suppose the semester averages from a college course are normally distributed with a mean of 70 and a standard deviation of 10. The notation would look like:

X~N(70,10)

X means the scores.

~ means "is distributed."

N means normal

(μ,σ) is (mean, standard deviation)

Draw the distribution represented by X~N(70,10)

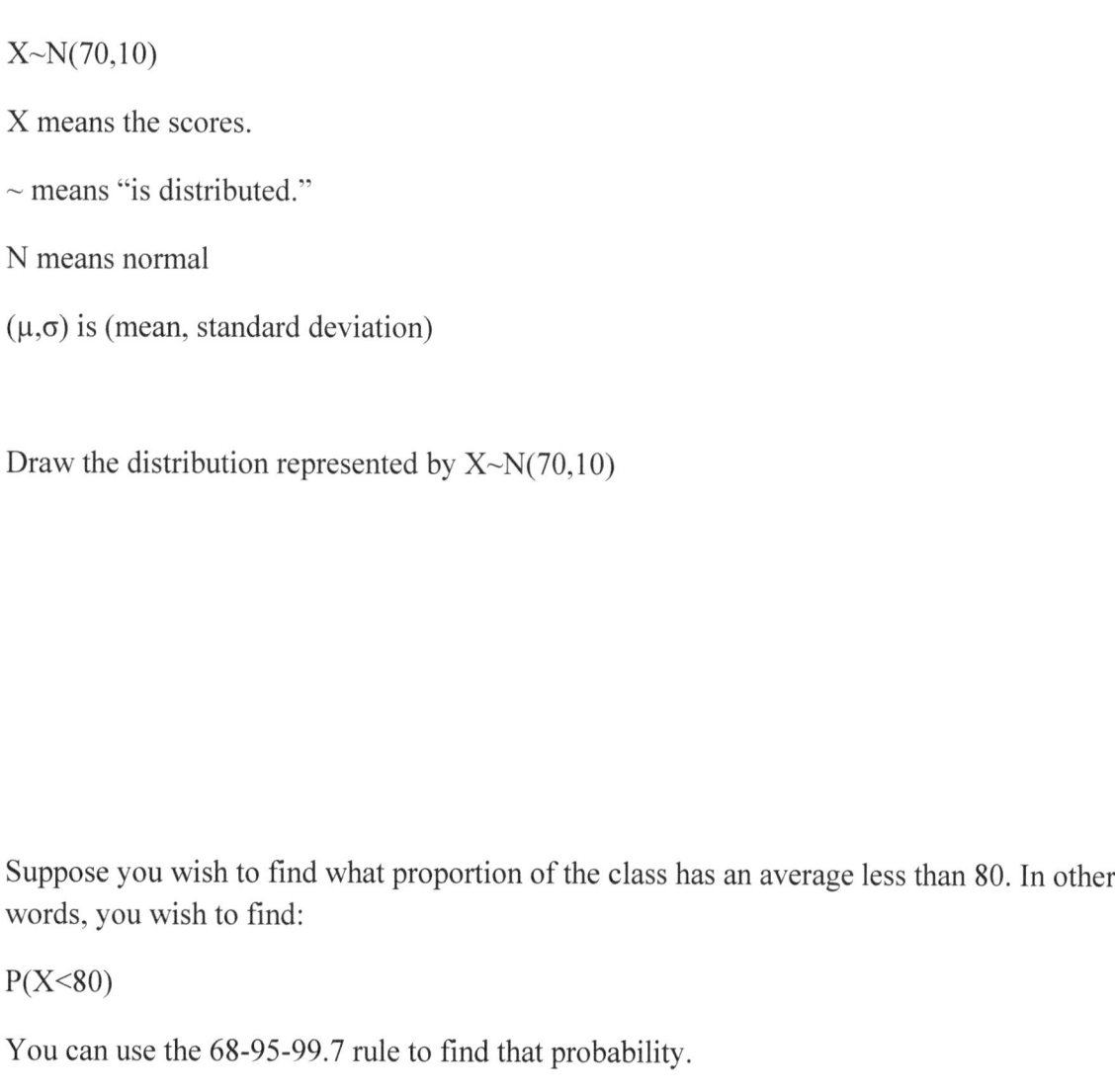

Suppose you wish to find what proportion of the class has an average less than 80. In other words, you wish to find:

P(X<80)

You can use the 68-95-99.7 rule to find that probability.

This rule will only be effective if you wish to find the proportion of the class who scored less than (or greater than) one of the seven values labeled on the bottom of the curve. If you wish to find out what proportion of the class had an average below 87, you would need another procedure.

Percentile:

Standardized score (Z-score):

Standardized scores can be shown by drawing the standard normal curve, which is:

$X \sim N(0,1)$

Referring to the curve that illustrated semester averages, $X \sim N(\mu, \sigma)$, you can determine where the value in question would fall on the standard normal curve. For instance, in the previous example we wished to find what proportion fell under 80. In other words, we wanted to find:

$P(X < 80)$

On the standard normal curve, that would be the same as:

$P(Z < 1)$

How to use the Standard Normal Table

First you need to write your Z score rounded to the nearest hundredth. (I call it "dollar and cents" form.)

The Z-score in the example is 1, so it needs to be written as:

Z = 1.00

Look at the ones and tenths places only (dollars and dimes), find those values on the left-most column of the table.

Then find the hundredths (pennies) on the top row. Find where those values intersect.

(Refer to second digit only.)

Z	.00	.01	.02	.03	.04......
.					
.					
0.8					
0.9					
1.0	.8413				
1.1					

Using the 68-95-99.7 rule, we had approximated P(X<80) to be .84. Using the table, we have found P(X<80) to be .8413.

Given any value on the standard normal table, you can find the percentile (proportion BELOW) on the table.

Find the percentiles for the given Z-scores.

P(Z< 2.45)   P(Z< -1.07)   P(Z< 0.03)

Those table values always tell you what proportion is below a given value. In order to find what proportion is ABOVE a given value, subtract the proportion found in the table from 1.

Find P(Z> -0.52)

By converting a given value, X, to a standardized score, Z, you can then use the table to determine your percentile.

Converting an actual value, X, into a Z

Suppose you were only given the following information:

X~N(70,10)

And you wish to find P(X<90)

In other words:

Mean = 70

Standard deviation = 10

The score in question is 90

Just using the numbers, can you determine how many standard deviations 90 falls from the mean?

Create a formula:

Converting actual values to percentiles in the TI83/84

Hit [2nd] [VARS] to get the DISTR feature.

Select option 2, normalcdf(

FOR THE PROBABILITY BELOW A GIVEN VALUE X:

(Lower bound, Upper bound, Mean, Standard deviation)

Your bounds will be from negative infinity to x. Negative infinity is typed in the calculator as   **-1E99,** which is found by typing:

-1 then **[2nd][,] to get the EE function,** then **99.**

Your calculator should look like:

Normalcdf(-1E99, X, μ, σ)

Press [ENTER]

FOR THE PROBABILITY ABOVE A GIVEN VALUE X

(Lower bound, Upper bound, Mean, Standard deviation)

Your bounds will be from x to infinity. Infinity is typed in the calculator as **1E99,** which is found by typing:

1 then **[2nd][,] to get the EE function,** then **99.**

Your calculator should look like:

Normalcdf(X, 1E99, μ, σ)

Press [ENTER]

Examples:

1. The heights of plants after 1 year of growth are normally distributed with a mean of 5.4 inches and a standard deviation of .8 inches.

What percent of plants will be shorter than 5 inches?

What percent of plants will be taller than 6.5 inches?

2. The incomes at a particular company are normally distributed with a mean of $43,850 and a standard deviation of $5000.

What percent of the employees have incomes under $30,000?

What percent earn more than $47,000?

What percent of employees earn between $40,000 and $50,000.

Percent to Actual

Suppose the scores for a college entrance exam are normally distributed such that $X \sim N(200,10)$. In order to be qualified for the honors college, you have to score in the 97.72nd percentile on the exam. What score marks the cutoff for eligibility?

Draw and label the normal curve.

This time we are reversing the direction of the last lesson. The last step in going from actuals to percentiles was to look up the z-score and find the percentile in the body of the table. Now you need to find the percentile in the body of the table and find the corresponding z-score by referring to the left and upper margins.

What z-score corresponds with the 97,72nd percentile?

By plotting that on the normal curve, you can see right away that the desired score is a 220. However, could you use just the numbers available (and no picture) to arrive at the same conclusion?

Mean = 200
Std. dev = 10
z-score (number of standard deviations from the mean) = 2.

What is the formula for converting a z-score to an actual value?

Now suppose your goal in the previous example isn't to get into the honors college, but rather just the college. If the college will only consider students who scored at or above the 30th percentile, what score would be the cutoff?

If you convert the 30th percentile into decimal form, you get .30. Since the values in the body of the table are written as 4-digit decimals, you need to add two zeroes to that decimal to make it .3000. However, you will not find .3000 in the body of the table, so you need to find the closest value you can, which is .3015. That corresponds to a z-score of -0.52.

Use the formula to complete the problem.

Suppose the requirements for the honors college were that the student achieve in the top 10%. What percentile would you be at if you scored in the top 10%?

YOU ALWAYS LOOK UP THE PERCENTILE (PERCENT BELOW YOU) IN THE TABLE.

The way to determine your percentile is to draw a quick sketch of a normal curve (numbers and labels aren't necessary) and draw a line where the cutoff is. Label the left and right sides of the cutoff. YOU ALWAYS LOOK UP THE VALUE TO THE **LEFT** OF THE CUTOFF, even if the 'desired' region is to the right. The table is a list of percentiles, which is always defined as the percent **below** you, which will always be the region to the left.

How to convert from percentiles to actual values on the TI 83/84

Hit [2ⁿᵈ] [VARS] to get the DISTR feature.

Select option 3, invNorm(

Enter the proportion BELOW the given value, the mean, and the standard deviation.

Examples:

The average age of the local semi-pro football team is 27.8 with a standard deviation of 1.4 years.

70% of the players are below what age?

What age marks the highest 10%?

SOL scores from a local school are normally distributed with a mean of 36.4 and a standard deviation of 3.1884.

If 91.6% of the school passed the SOL, what score marks the cutoff for passing?

If the school wanted to set up a class for the lowest 14% of students, what score is the cutoff?

Z-score transformation practice

1. College entrance exam scores are distributed such that X~N(250, 25).

a) If the college will only consider accepting people who score above the 35th percentile, what score would mark the cut-off for consideration?

b) You scored a 267. At what percentile do you fall?

c) Your friend scored a 239. What percent of the people taking this exam scored better than her?

d) The top 10% of people can be considered for the honors college. What score marks the cut-off for consideration?

e) What percent of the people taking this exam score between a 220 and a 260?

2. **IQs are normally distributed with a mean of 100 and a standard deviation of 15.**

a) If your IQ is 107, what percent of people have an IQ above you?

b) Find $P(90 < X < 110)$. What does that mean?

c) A school wants to provide remediation for students with the lowest 14% of IQs. What score marks the cutoff?

d) The same school wants to provide advanced placement for students with the top 8% of IQs. What score marks the cutoff?

e) Ben's IQ is 97. What is his percentile?

Sampling Distributions

What is the expected value for the mean roll if a fair, six-sided die is rolled numerous times?

X	1	2	3	4	5	6
P(X)						

Expected Value: **Standard deviation:**

If each of you are tasked with rolling a die 20 times and averaging your outcomes together, will you all be guaranteed to get exactly that amount as your average?

Roll your die 20 times, record the outcomes and average them together.

Roll #	Outcome	Roll #	Outcome	Roll #	Outcome	Roll #	Outcome
1		6		11		16	
2		7		12		17	
3		8		13		18	
4		9		14		19	
5		10		15		20	

Your average =

Since your average came from the sample, your value is denoted as $\overline{X}$.

I can predict what your sample averages, $\bar{x}$, will come back looking like. While I can't predict exactly what YOU will get, I have a good idea of what the class's values will be.

Repeated sample averages from samples of the same size from the same population will be approximately distributed:

$$N \sim (\mu, \frac{\sigma}{\sqrt{n}})$$

In this case, $\mu = 3.5$, $\sigma = 1.707825$, and your sample size $n = 20$.

Therefore, I can predict your values will come back approximately $N \sim (3.5, 0.3819)$

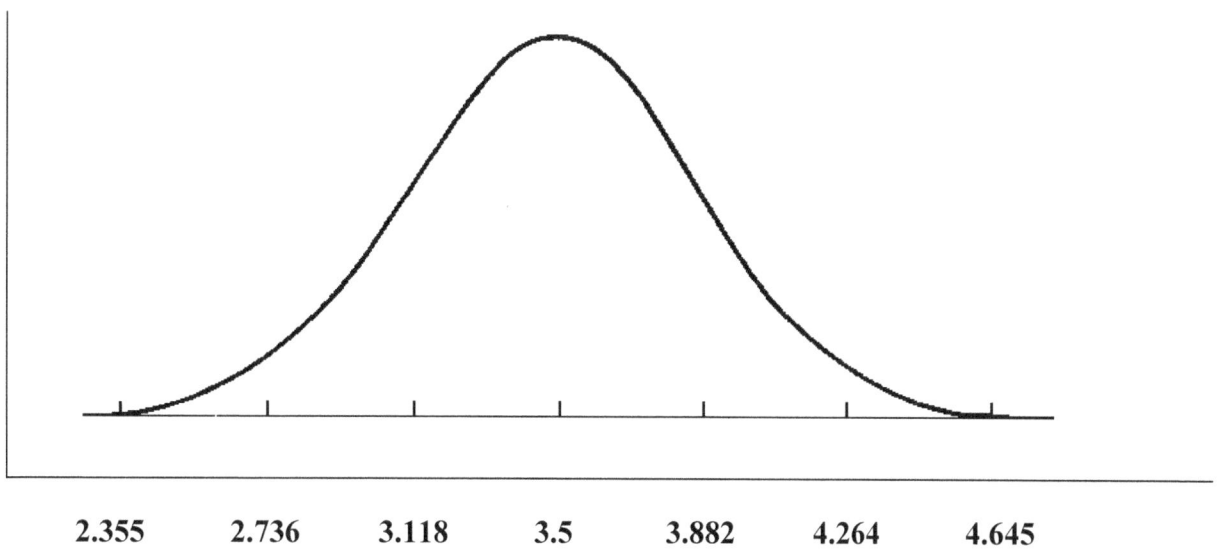

| 2.355 | 2.736 | 3.118 | 3.5 | 3.882 | 4.264 | 4.645 |

In other words, I can expect 68% of you to report sample averages that fall between…

95% of you should report sample averages between…

99.7% of you should report sample averages between…

81

Assumptions:

Suppose I asked you to go out tonight and acquire a sample of men and determine two separate sample averages…their height and their weight. Your sample averages should come back fitting a normal curve, but only IF TWO THINGS ARE TRUE OF YOUR SAMPLES.

Rule number 1:

Rule number 2:

Example:

Baby weights are such that X~N(7.1, 1.8). Draw and label the normal curve showing the weights of individual babies:

What is the probability that an individual baby will weigh less than 6.6 pounds?

Suppose I task each of you with going out tonight and getting a random sample of 10 babies and averaging their birth weights together.

Are the assumptions for normality met?

If so, what would the sampling distribution look like? Draw that curve.

What is the probability that the AVERAGE weight from a sample of 10 babies will be less than 6.6 pounds?

The formulas are as follows:

For individuals:

$$Z = \frac{x - \mu}{\sigma}$$

For sample means:

$$Z = \frac{\bar{x} - \mu}{\sigma / \sqrt{n}}$$

Example:

For those participating in a diet program, the weight losses for last year are X~N(23.2, 8.1).

What is the probability that a randomly selected dieter will have lost more than 25 pounds?

What is the sampling distribution for random samples of size 40?

What is the probability that the average weight loss from 40 randomly selected dieters will be more than 25 pounds?

To find the probabilities on the TI 83/84:

For individuals:

Hit [STAT] and arrow over twice to TESTS.

Select option 1 and hit [ENTER]

Input: Stats

μ_0 = population mean

σ = population standard deviation

$\bar{x}$ = the individual value

n = 1

Determine if you want $\neq$, < or >

Select calculate

For means:

Hit [STAT] and arrow over twice to TESTS.

Select option 1 and hit [ENTER]

Input: Stats

μ_0 = population mean

σ = population standard deviation

$\bar{x}$ = the sample mean

n = the sample size

Determine if you want $\neq$, < or >

Select calculate

Examples:

1. Salaries at a particular company are distributed such that the mean is $54,845 and the standard deviation is $8,255.

   a) Find the probability that a randomly selected employee will earn more than $50,000.

   b) Find the probability that a randomly selected group of 70 employees will have an average salary over $56,000.

   c) Find the probability that a randomly selected group of 60 employees will have an average salary between $53,000 and $57,000.

   d) Explain why you wouldn't want to use this data to find the probability that a randomly selected group of 10 people would earn less than $52,750.

   e) What is the probability that a randomly selected employee would earn less than $54,000? What is the probability that a randomly selected group of 45 employees would earn less than $54,000?

   f) What does the sampling distribution look like for random samples of size 45?

Sampling Distributions for Proportions

Let π represent the proportion of the population who fits a certain description. Suppose I task each of you with going out and getting a sample of n subjects and determining what proportion of those subjects fits the given description. You are not all guaranteed to come back with a sample proportion that matches π exactly, but provided your samples meet certain assumptions, the values should come back normally distributed about π. The standard deviation of that curve can be found by determining:

$$\sqrt{\frac{\pi(1-\pi)}{n}}$$

Therefore, the sample proportions you come back with should be approximately distributed $p \sim N(\pi, \sqrt{\frac{\pi(1-\pi)}{n}})$

Assumptions:

This will be true if two things are true of your samples:

Rule #1:

Rule #2:

Examples:

7.6% of VCU students are over the age of 30. What would be the sampling distribution if repeated random samples of size 50 are taken?

If 67.2% of all VCU students vote, what can be the expected sampling distribution if repeated random samples of size 100 are taken?

20% of people are left-handed. What would be the sampling distribution if repeated random samples of size 100 were taken? Draw and label the curve.

What would be the probability of observing a random sample of size 100 where less than 17% of the sample was left-handed?

Using the same example, what would be the probability of observing a random sample of size 100 in which more than 22% of the sample was left-handed?

Determining probabilities with sampling distributions on proportions on the TI 83/84

Press [STAT] and arrow over twice to TESTS.

Select option 5: 1-PropZTest...

For P_0: Enter the value for π

For x: Enter the number of successes in your sample

For n: Enter the sample size

Determine if you wish to find the proportion above or below.

Select Calculate

Examples:

27.5% of VCU students have a stat course required for their major. If you were to take a random sample of 200 VCU students, what is the probability that less than 27% have stats required for their major?

16.2% of teens have tried smoking. What is the probability that a randomly selected group of 300 teens would reveal that more than 20% have tried smoking? What is the sampling distribution?

Limitation to the TI 83/84 calculator:

UNIT 3

Confidence Intervals on Proportions

Step 1. Determine p.

    a) could be given in the form of a decimal
    b) could be given in the form of a percent
    c) could be given as an "out of" statement, which would require a division (y/n)

Step 2. Check assumptions:
    Sample must be random
    $np \geq 10$, $n(1\text{-}p) \geq 10$

Step 3. Compute standard deviation.

Plug the value for p into $\sqrt{\dfrac{p(1-p)}{n}}$, where n is the sample size

Step 4. Determine Z*

    Look up the percent confidence on the Z* chart and choose the appropriate Z* value.

Step 5. Plug those values into the formula

$$p \mp Z * \sqrt{\dfrac{p(1-p)}{n}}$$

You will get two values, the lower end of the interval and the upper end of the interval

Confidence interval instruction

A random sample of 100 men revealed that 20% would seek treatment if they encountered hair loss. Find a 95% confidence interval for the proportion of men who would seek treatment for hair loss.

1. Find the sample proportion p.

2. Check the assumptions.

3. If the assumptions are met, compute the standard deviation by using the formula. Once you have a p value and a standard deviation value, you can complete a normal curve. (You know it's a normal curve because of the test in step 2.) Put the p value in the middle of the curve and increase/decrease by the standard deviation value.

You can be roughly 95% confident that between _____ and _____ percent of all men would seek treatment for hair loss.

What about 68% confident?

Z* tells you how many standard deviations you need to go in each direction to achieve the desired confidence level.

Suppose you want to determine the approximate 95% confidence interval, but you don't have the picture. The only pieces of information you have are:

$p = .20$
std. dev. $= .04$
95% confidence $\rightarrow Z^* = 2$

Use that information to arrive at the formula:

6. Make a confidence statement, which has the following three parts:

a) state how confident you are
b) tie it to the problem, making sure to reference "all" or "population"
c) state the interval

Confidence interval problem 2

2. A recent study of 200 randomly selected pregnant women showed that 140 had side effects from the iron supplements. Find a 99% confidence interval for the proportion of all pregnant women who suffer side effects from iron supplements.

Show if the assumptions are met.

If appropriate, calculate the 99% confidence interval

Make a confidence statement

Finding a confidence interval for a proportion on the TI 83/84

Press [STAT] and arrow over twice for TESTS.
Select A: 1-PropZInt
x: number of successes
n: sample size
C-Level: desired confidence level
Select Calculate

In class example for confidence intervals

1. In a random sample of 100 VCU students, 38% said that stats is required for their major.

Are the assumptions for normality met? State why or why not.

If appropriate, calculate and interpret a 90% confidence interval for the proportion of all stats students for whom stats is a required course.

2. A random sample of 32 students showed that 6.25% have admitted to cheating.

Are the assumptions for normality met? State why or why not.

If appropriate, calculate and interpret a 95% confidence interval for the proportion of all students who have admitted to cheating.

3. A random sample of 1000 shoppers showed that 880 are more likely to shop if there is a store-wide sale.

Are the assumptions for normality met? State why or why not.

If appropriate, calculate and interpret a 99% confidence interval for the proportion of all shoppers who are more likely to shop during a store-wide sale.

What does a confidence interval mean?

Penny activity:

Flip your penny 20 times and determine what proportion of the time "heads" appeared. You may use the table to record your results.

Flip	Heads?	Flip	Heads?	Flip	Heads?	Flip	Heads?
1		6		11		16	
2		7		12		17	
3		8		13		18	
4		9		14		19	
5		10		15		20	

What is the value of your sample proportion $\hat{p}$?

Can you say conclusively that this has to be the population proportion p?

Check the assumptions:

Calculate a 95% confidence interval for the true proportion (p) of all coin flips that will land heads up.

Make a confidence statement.

What percent of your calculated intervals should contain the true proportion of 0.5? What percent should miss?

97

Finding confidence intervals on means, sigma known

Step 1. Determine $\overline{X}$ **, the average from the sample. (Often given)**

Step 2. Determine if the assumptions for normality are met.

    a) **Random sample**

    b) **Normally distributed population**
           OR
      Sample larger than 40

Step 3. Determine standard deviation, σ **, and sample size, n.**

Step 4. Determine Z*

Look up the percent confidence on the Z* chart and choose the appropriate Z* value.

Step 5. Plug those values into the formula

$$\overline{X} \pm z^*\left(\frac{\sigma}{\sqrt{n}}\right)$$

You will get two values, the lower end of the interval and the upper end of the interval

Confidence Intervals for Means with Sigma known

Factory machines are supposed to produce products that are the correct size, with a very small margin of error. If the products coming off the assembly line are too large or too small, they will often not be able to work properly. With consistent use, machines often become "off," and the resulting products do not come out the correct size. For this reason, the quality control staff will periodically take a random sample of products coming from a particular machine and determine if the products are of an acceptable size.

In this instance, the size of the products is normally distributed and the standard deviation of the machine is often already known.

1. In question is the size of the washers being produced by an aging machine. 20 washers are randomly selected and measured, and their average diameter is 14.8 mm. The standard deviation from this machine is known to be .03 mm, and the sizes of the washers are normally distributed.

Are the assumptions for normality met? Is sigma known?

Compute and interpret a 95% confidence interval for the average diameter of a washer being produced by this machine.

If the washers are supposed to have an average of 15 mm, is this machine in need of repair?

Finding confidence intervals on means with sigma known on the TI 83/84

Press [STAT] and arrow over twice to TESTS
Choose option 7: ZInterval
Select Stats to input the mean and standard deviation
σ: Enter standard deviation
$\bar{x}$: sample mean
n: sample size
C-level: state the confidence level
Select Calculate

2. 50 buttons are randomly selected from a machine, and their average diameter is 5.98 mm. That machine is known to produce buttons that have normally distributed sizes with a standard deviation of .08 mm.

Are the assumptions for normality met? Is sigma known?

Create and interpret a 99% confidence interval for the average button diameter from that machine.

If the average button diameter is supposed to be 6 mm, does the machine need attention?

Finding confidence intervals on means with sigma known on the TI 83/84

Press [STAT] and [ENTER] to put the information into a list
Press [STAT] and arrow over twice to TESTS
Choose option 7: ZInterval
Select Data
σ: Enter standard deviation
List: Choose which list contains the data
Freq: 1
C-level: state the confidence level
Select Calculate

3. A random sample of ten 2X4's were measured, and their actual widths, in inches, were

4.06   4.01   3.98   3.99   4.10   4.01   3.97   4.02   3.99   4.05

The machine that creates these boards is known to have a standard deviation of .04 and produce normally distributed widths.

Are the assumptions for normality met? Is sigma known?

Create and interpret a 90% confidence interval for the average width of the boards coming off the machine.

If they are truly supposed to be 4 inches wide, does the machine need to be examined?

Finding confidence intervals on means, sigma unknown

Step 1. **Determine** $\overline{X}$ **, the average from the sample. (Often given)**

Step 2. Check assumptions (Random only)

Step 3. Determine sample standard error, s, and sample size, n.

Step 4. Determine t*

Find the degrees of freedom by subtracting one from the sample size

Look up the percent confidence on the t* chart and choose the appropriate t* value.

Step 5. Plug those values into the formula

$$\overline{X} \pm t^* \left(\frac{s}{\sqrt{n}} \right)$$

You will get two values, the lower end of the interval and the upper end of the interval

The potential danger of estimating σ with s.

Assume the national average for the birth weight of babies is 7 pounds, but you don't know this.

Suppose you want to find a confidence interval for the mean birth weight of all babies, but you didn't know the population standard deviation. In order to create this confidence interval, you use a sample of 4 babies, and their weights are as follows:

7.1 lbs
7.3 lbs
7.0 lbs
7.2 lbs

Find $\bar{x}$ and s.

Is $\bar{x}$ a good estimate for the population average of 7 pounds?

Draw and label the normal curve for the mean birth weight of babies.

You can be roughly 95% confident that the average birth weight for all babies is between:

In class assignment for confidence intervals on means with Sigma unknown

1. The weights of newborn babies are normally distributed in the population. You have randomly selected 10 babies to be in a study, and their birth weights (in pounds) are as follows:

7.3   6.8   8.3   5.2   9.4   6.7   7.2   7.1   7.7   6.2

If appropriate, calculate and interpret a 90% confidence interval for the average birth weight of all babies.

Finding confidence intervals on means with sigma unknown on the TI 83/84

Press [STAT] and arrow over twice to TESTS
Choose option 8: TInterval
Select Stats to input the mean and standard error
$\overline{x}$**: sample mean**
sx: standard deviation from the sample
n: sample size
C-level: state the confidence level
Select Calculate

2. The times, in seconds, of 51 randomly selected elementary school students running the 100 yard dash were recorded, and the average time was 26.3 seconds. The standard deviation from the sample was 3.8 seconds. If appropriate, find and interpret a 95% confidence interval for the average 100 yard dash time for all students at the school.

Why random is the only assumption when using t*.

Finding confidence intervals on means with sigma unknown on the TI 83/84

Press [STAT] and [ENTER] to put the information into a list
Press [STAT] and arrow over twice to TESTS
Choose option 7: ZInterval
Select Data
σ: Enter standard deviation
List: Choose which list contains the data
Freq: 1
C-level: state the confidence level
Select Calculate

3. The weights of 10 randomly selected adults were recorded as follows:

163   189   125   258   164   192   201   310   142   187

If appropriate, calculate and interpret a 99% confidence interval for the average weight for all adults.

Confidence Interval Review

1. The quality control staff is concerned that a machine making bicycle wheel spokes has begun creating spokes that are too long. The lengths of the spokes produced are normally distributed, and the known standard deviation of the machine is .12 inches. A random sample of 10 spokes was collected, and their lengths were as follows:

9.75   9.76   9.68   9.7   9.72   9.73   9.74   9.75   9.7   9.71

If appropriate, calculate and interpret a 95% confidence interval for the average length of spokes produced from this machine.

If the average is supposed to be 9.7 inches, is this machine in need of repair? Explain.

2. A random sample of 150 post-menopausal women showed that 96 showed signs of decreased bone density. If appropriate, find and interpret a 90% confidence interval for the proportion of all post-menopausal women who have decreased bone density.

3. A random sample of 61 people took an experimental weight loss drug. After 6 months, their average weight loss was 23.6 pounds with a standard deviation of 5.4 pounds. If appropriate, find and interpret a 99% confidence interval for the average weight loss for all people who will take this drug.

Comparing proportions using confidence intervals

Sometimes the goal is to determine how different two sample proportions are, and this can be done using a confidence interval for the difference of the two proportions.

Suppose we wish to see if there is a difference in the proportion of students who receive A's in two professors' classes. Random samples of size 100 were taken from each professor, and the proportions of the students who received A's were recorded. Professor Smith's sample revealed that 30 students received an A, whereas Professor Jones's sample showed that number to be 20.

Step 1: Organize the information.

Step 2: Check the assumptions.

Step 3: Determine the difference between your two sample proportions. ($p_1 - p_2$)

Step 4: Place that value in the middle of a normal curve (since the assumptions for normality were met.) Complete the normal curve by calculating the standard error of the difference:

$$SE = \sqrt{\frac{(p_1)(1 - p_1)}{n_1} + \frac{(p_2)(1 - p_2)}{n_2}}$$

Complete the normal curve.

Calculate your confidence interval based on the desired level of confidence. If your interval contains zero, there is a possibility the population proportions are the same.

$$(p_1 - p_2) \mp z* \sqrt{\frac{(p_1)(1-p_1)}{n_1} + \frac{(p_2)(1-p_2)}{n_2}}$$

Example:

You wish to see if there is a difference in the proportion of students receiving free lunch at Elementary school A compared to Elementary school B. A random sample of 150 students at School A revealed that 17.333% of the students received free lunch, and a random sample of 120 students at School B revealed that 12.5% received free lunch. Does this show at 95% confidence that the proportions of students receiving free lunch are different for the two schools?

Computing this interval on the TI 83/84:

Select [STAT] and arrow over twice to TESTS.

Choose option B: 2-PropZInt…

x1: The number of successes in sample 1
n1: The sample size of sample 1
x2: The number of successes in sample 2
n2: The sample size of sample 2

Select the confidence level in C-Level
Select Calculate

If the interval contains zero, we failed to show the proportions were different at the given confidence level.

Example: You wish to determine if the rate of dissatisfied customers is different at Store A than Store B. In a random sample of 200 shoppers at Store A, 18 said they were dissatisfied. A random sample of 160 people from Store B showed that 11 were dissatisfied. What is your conclusion? (Be 95% confident.)

2. You want to see what proportion of students have had their car towed at the MCV campus as opposed to Monroe Park campus. 100 randomly selected students from each campus revealed that 8% of students at Monroe Park have had their car towed, whereas 6% of students at MCV have been towed. What is your conclusion? (Be 90% confident.)

Comparing means using confidence intervals
Sigma known

Using the same convention as before, we wish to see if the confidence interval for the difference of the two population means contains zero. This time the formula we use is as follows:

$$\left(\bar{x}_1 - \bar{x}_2\right) \mp z^* \sqrt{\left(\frac{\sigma_1^2}{n_1} + \frac{\sigma_2^2}{n_2}\right)}$$

Where $\left(\bar{x}_1 - \bar{x}_2\right)$ represents the difference between the two sample means, and

$\sqrt{\left(\frac{\sigma_1^2}{n_1} + \frac{\sigma_2^2}{n_2}\right)}$ represents the standard deviation of the difference of the means.

Example:

IQs are normally distributed in the population with a known standard deviation of 15 points. Two schools are being compared to see if there is a difference in the average IQ of those attending. Gleason Elementary had a random sample of 15 students with an average IQ of 103.8, and Riverbrook Elementary had a random sample of 18 students with an average IQ of 99.4. Does this show, at 95% confidence, that the schools have different average IQs?

Perform that same example using the TI 83/84

Determining a confidence level for a difference of two means (with sigma known) on the TI 83/84

Press [STAT] and arrow over twice to TESTS
Choose option 9: 2-SampZInt
Choose Stats
$\sigma 1$**: Enter the known standard deviation for sample 1**
$\sigma 2$**: Enter the known standard deviation for sample 2**
$\bar{x}1$**: Enter the sample average for sample 1**
n1: Enter the sample size for sample 1
$\bar{x}2$**: Enter the sample average for sample 2**
n2: Enter the sample size for sample 2
C-Level: Enter the desired confidence level
Select calculate.

Examples:

SAT scores have a known standard distribution of 312.4 points. Two schools are being compared to see if their average SAT scores are different. Ten students from school A were randomly selected, and their average SAT score was 1516.7. Twelve students from school B were randomly selected and their average SAT score was 1501.3. Does this show, at 90% confidence, that the two schools have different average SAT scores?

Two machines produce buttons that have a known standard deviation of .08 mm. Of interest is if the machines have been producing buttons of the same average size. A SRS of size 40 from machine A revealed an average diameter of 15.02 mm, and a SRS of size 50 from machine B revealed an average diameter of 14.89 mm. Does this show, at 90% confidence, that the machines have different average diameters?

Comparing means using confidence intervals
Sigma unknown

Using the same convention as before, we wish to see if the confidence interval for the difference of the two population means contains zero. This time the formula we use is as follows:

$$\left(\bar{x}_1 - \bar{x}_2\right) \mp t^* \sqrt{\left(\frac{s_1^2}{n_1} + \frac{s_2^2}{n_2}\right)}$$

Where $\left(\bar{x}_1 - \bar{x}_2\right)$ represents the difference between the two sample means, and

$\sqrt{\left(\frac{s_1^2}{n_1} + \frac{s_2^2}{n_2}\right)}$ represents the sample standard deviation of the difference of the means.

Example:

A company that makes spark plugs wishes to see if there is a difference in the average life expectancy of their spark plugs compared to a competitor. A simple random sample of 55 spark plugs from their own company revealed an average lifetime of 121,543 miles with a standard deviation of 6,437 miles. The competitor's simple random sample of 42 spark plugs showed an average lifetime of 117,460 miles with a standard deviation of 8,756 miles. Does this show, at 95% confidence, that there is a difference in the average life expectancy of the two brands?

Determining a confidence level for a difference of two means (with sigma unknown) on the TI 83/84

Press [STAT] and arrow over twice to TESTS
Choose option 0: 2-SampTInt
Choose Stats
$\bar{x}$**1: Enter the sample average for sample 1**
sx**1: Enter the standard deviation from sample 1**
n1: Enter the sample size for sample 1
$\bar{x}$**2: Enter the sample average for sample 2**
sx**2: Enter the standard deviation from sample 2**
n2: Enter the sample size for sample 2
C-Level: Enter the desired confidence level
Pooled: select No
Select calculate.

Mixed problems:

1. A researcher wants to determine if there is a difference in the average number of puppies born to dogs of two different breeds. A random sample of 40 Dalmatian litters revealed an average of 5.1 puppies with a standard deviation of 1.8. A random sample of 38 Beagle litters had an average of 4.6 puppies with a standard deviation of 1.5. Does this provide evidence (at 95% confidence) that the dog breeds have different average numbers of puppies per litter?

2. A bandage company wants to see how long their bandage will stick on a child's hands under extreme play conditions. A simple random sample of 65 children placed bandages on their hands and went out to a playground. The mean amount of time the bandages stayed on was 4.5 hours with a standard deviation of 1.2 hours. Calculate and interpret a 90% confidence interval for the mean amount of time a bandage will stay on.

3. A pharmaceutical company wants to determine what proportion of people who have gotten the flu shot will still get the flu. A random sample of 180 people with the flu shot was polled, and 5% of them got the flu. Calculate and interpret a 99% confidence interval for the proportion of all people who will get the flu after a flu shot.

4. An independent research organization wants to determine if there is a difference in the proportion of people who get cavities with two popular toothpastes. A random sample of 300 people who use Toothpaste A revealed that 21 had cavities, while a random sample of 350 people who use Toothpaste B showed that 22 had cavities. Does this show, at 95% confidence, that the two proportions are different?

5. A quality control person wants to see if the average size of the parts from an aging machine has shifted. The machine is known to produce pieces that are normally distributed and have a known standard deviation of 0.52 centimeters. A random sample of 15 parts showed an average length of 15.3 centimeters. Calculate and interpret a 99% confidence interval for the average length of all parts coming from this machine. If the machine is supposed to produce parts that have an average length of exactly 15 centimeters, does this machine need repair?

Hypothesis testing on proportions

Step 1. Write the null and alternative hypotheses

Null: Always contains an = sign and CONTAINS THE VALUE FOR "π"
States the current belief in the population

Alternative: Always contains an <,>, or ≠
States what the researcher believes will be true.

Step 2. Determine p BASED ON THE STUDY

P will be given as a decimal, percent, or "out of" statement
THEN CHECK THE ASSUMPTIONS

Step 3. Plug π, n and p into the formula:

$$Z = \frac{p - \pi}{\sqrt{\dfrac{\pi(1 - \pi)}{n}}}$$

Always use the value from the null to determine the standard deviation in the denominator.

***THIS FORMULA FOLLOWS THE FAMILIAR FORM

$$Z = \frac{\text{OBSERVED SCORE} - \text{MEAN}}{\text{STD. DEV}}$$

Step 4. Look up the Z-score on the chart to find the associated percentile

Step 5. Find the P-value

If the alternative hypothesis in step 1 is a "less than" statement, the percentile is the p-value.

If the alternative hypothesis in step 1 is a "greater than" statement, you must subtract the percentile from 1 to get the p-value

If the alternative hypothesis in step 1 is a "not equal to" statement, then look up the negative version of the Z-score and double the percentile to get the p-value.

117

Step 6. Draw the conclusion

If the p-value is less than the significance level given in the problem, then reject the null in favor of the alternative. Acknowledge you may have made a type I error.

If the p-value is greater than the significance level given in the problem, then fail to reject the null. Acknowledge you may have made a type II error.

Step 7. Make concluding statement

If you reject the null, state there is sufficient evidence to support the alternative (expressly stating the alternative.)

If you fail to reject the null, state there is insufficient evidence to support the alternative (stating the alternative.)

In class example for hypothesis testing on proportions

A study done in 1980 revealed that 50% of teens aged 16-19 were sexually active. You believe the current percentage is higher than that. You plan to randomly select 100 teens in that age bracket and determine if they are sexually active.

You must begin under the assumption that the previous finding is still true unless you can prove otherwise. If that's the case, 50% of the population would still be sexually active today (notice that value would be π and not p). You should put that value in the center of a normal curve. Find the value of the standard deviation.

Draw the appropriate normal curve.

If you conduct your survey and come up with a percentage that is in line with the values in the above normal curve, then your data supports the claim from 1980. If the percentage from your sample is not aligned with the values in the normal curve above, then your study is contradicting the study from 1980, and you have what is called a *significant finding*.

A study done in 1980 revealed that 50% of teens aged 16-19 were sexually active. You believe the current percentage is higher than that. You survey 100 randomly selected teens and discover that 61% are sexually active. Are these findings significant at the .05 level?

Null Hypothesis:

Alternative Hypothesis:

P:

Test statistic: This tells you how many standard deviations you are from the mean of the normal curve on the previous page.

P-value: This tells you how likely you would be to get your sample value if the 1980 information were still true. (What % chance do you have of getting a sample that is 61% sexually active if the actual population percentage is 50% sexually active?)

Conclusion: Is your finding significant? (Is your value far enough away from 50% to be considered relevant?) In other words, is your value beyond your significance level?

Possible error?

Make a concluding statement in terms of the alternative hypothesis.

Finding the test statistic and p-value (for hypothesis testing on proportions) on the TI 83/84

Press [STAT] and arrow over twice to TESTS
Select option 5: 1-PropZTest
P0: enter the value for π
X: enter the number of successes in the sample
N: enter the sample size
State whether the alternative was $\neq$, > or <
Select Calculate

Hypothesis Testing on Proportions Practice Examples

An existing drug has 85% effectiveness; you have come up with a new drug that you think is more effective. In a study you conducted, you found 182 out of 200 randomly selected people showed improvement on your new medication. Test the hypothesis at the significance level .01.

1. State the hypotheses:

2. Check the assumptions:

3. Find the value of the test statistic (Z-score.)

4. What is the p-value?

5. What conclusion do you draw? What type of error could you have made?

6. Make a concluding statement.

Example 2:

A researcher wants to show that fewer than 23% of teens smoke. In a random sample of 100 teens, 18% claim they are smokers. Test the hypothesis at the significance level .05.

1. State the hypotheses:

2. Check the assumptions:

3. Find the value of the test statistic (Z-score):

4. What is the p-value?

5. What conclusion do you draw? What type of error could you have made?

6. Make a concluding statement.

Hypothesis testing on means with sigma known

Step 1. Write the null and alternative hypotheses

    Null: Always contains an = sign and CONTAINS THE VALUE FOR "μ "
    Alternative: Always contains an <,>, or $\neq$

Step 2. Determine $\overline{X}$ and s BASED ON THE STUDY and check the assumptions.

Step 3. Plug $\overline{X}$, μ, σ and n into the formula

$$z = \frac{\overline{X} - \mu}{\sigma / \sqrt{n}}$$

Step 4. Look up the Z-score on the chart to find the associated percentile

Step 5. Find the P-value

    If the alternative hypothesis in step 1 is a "less than" statement, the percentile is the p-value.

    If the alternative hypothesis in step 1 is a "greater than" statement, you must subtract the percentile from 1 to get the p-value

    If the alternative hypothesis in step 1 is a "not equal to" statement, then look up the negative version of the Z-score and double the percentile to get the p-value.

Step 6. Draw the conclusion

If the p-value is less than the significance level given in the problem, then reject the null in favor of the alternative. Acknowledge you may have made a type I error.

If the p-value is greater than the significance level given in the problem, then fail to reject the null. Acknowledge you may have made a type II error.

Step 7. Make concluding statement

In class example for hypothesis testing on means with sigma known

The weights of newborn babies are normally distributed with a mean of 7 pounds and a known standard deviation of 2.5 pounds. You wish to see if a particular disease causes women to have, on average, smaller babies. You plan to take a random sample of 25 women with this disease to see if the average weight of their first born babies is less than 7 pounds, at the significance level .05.

Draw and label the normal curve for the average birth weight of 25 babies.

Your sample yielded an average of 6.4 pounds. Test to see if this finding is significant.

Hypotheses:

Assumptions:

Test statistic:

P-value:

Conclusion:

Concluding statement:

How to obtain the test statistic and p-value (for hypothesis testing on a single with sigma known) on the TI 83/84

Press [STAT] and then arrow over twice for TESTS
Choose option 1: Z-Test…
Select Stats
μ0: Population mean from the null
σ: Population standard deviation
$\bar{x}$: Sample mean
n: Sample size
State whether the alternative is a $\neq$, < or >
Select Calculate

Hypothesis Testing on Means Practice Examples

The heights of women in this country are normally distributed with a mean of 64 inches and a known standard deviation of 2.5 inches. You want to see if a group of women with "Medical condition A," on average, shorter than American women as a whole. You randomly select 17 women with this condition, and their average height is 62.7 inches. Is this finding statistically significant at the .01 level?

Hypotheses and assumptions:

Test Statistic:

P-Value:

Decision and concluding statement:

The average age of a woman at the birth of her first child 27, with a known standard deviation of 3. You believe that women with alcoholic fathers have a different average than 27, although you are not sure in which direction. You randomly select 100 women with alcoholic fathers and determine their average age at the birth of their first child is 27.8. Is this significant at the .05 level?

Hypotheses and assumptions:

Test Statistic:

P-Value:

Decision and concluding statement:

Hypothesis testing on means with sigma unknown

Step 1. Write the null and alternative hypotheses

>**Null:** Always contains an = sign and CONTAINS THE VALUE FOR " μ "
>**Alternative:** Always contains an <,>, or ≠

Step 2. Determine $\bar{X}$ and s BASED ON THE STUDY and check the assumptions.

Step 3. Plug $\bar{X}$, μ, s and n into the formula:

$$t = \frac{\bar{X} - \mu}{s/\sqrt{n}}$$

Step 4. Find the p-value…

Look in the row with the correct degrees of freedom for the problem.
Find the two values in the body of the table which surround the positive version of the t value you got in step 3
Look up the table to find the percentages which are at the heading of the two surrounding values.
If your alternative is a less than or a greater than statement, you can say that your p-value falls between those two percentages.
If your alternative is a not equal to statement, you must double those percentages to find the range of your p-value.

Step 6. Draw the conclusion

If the p-value is less than the significance level given in the problem, then reject the null in favor of the alternative. Acknowledge you may have made a type I error.

If the p-value is greater than the significance level given in the problem, then fail to reject the null. Acknowledge you may have made a type II error.

Step 7. Make your concluding statement.

In class example, hypothesis testing on means, sigma unknown

You think the new medication "B" will improve blood pressure by more than 14 points. A random sample of 101 patients had an average improvement of 14.8 points after using medication B. The standard deviation from the sample was 3.6 points.

Hypotheses and assumptions:

Test Statistic:

P-Value:

Decision and concluding statement:

How to obtain the test statistic and p-value (for hypothesis testing on a single with sigma unknown) on the TI 83/84

Press [STAT] and then arrow over twice for TESTS
Choose option 2: T-Test…
Select Stats
μ**0: Population mean from the null**
$\bar{x}$**: Sample mean**
Sx: Standard deviation from the sample
n: Sample size
State whether the alternative is a $\neq$, < or >
Select Calculate

Hypothesis Testing on Means Practice Examples
Sigma Unknown

A researcher claims that the *average* IQ will be above 100 upon completion of an educational program. He places 41 randomly selected students in the program, and in the end their average IQ is 103 with a standard deviation of 15. Is this finding statistically significant at the alpha = .05 level?

Hypotheses and assumptions:

Test Statistic:

P-Value:

Decision and concluding statement:

You believe the average cholesterol level will be lower on a new exercise program. The current average level is 187.2. In your random sample of 100 people, the average was 185.1 with a standard deviation of 8.2. Is this significant at the .05 level?

Hypotheses and assumptions:

Test Statistic:

P-Value:

Decision and concluding statement:

Mixed problems:

1. You want to see if "medical condition A" causes a difference in the average length of a woman's menstrual cycle, although you are not sure if it would become longer or shorter. The lengths menstrual cycles are normally distributed with an average of 28 days, with a known standard deviation of 2.2 days. Seventeen randomly selected women have an average length of 29.4 days. Are these findings significant at the .02 level?

2. A recent study declared that 30% of fourth graders perform below grade level in mathematics. You think the percentage is lower than that. A random sample of 300 fourth graders showed that 81 performed below grade level. Is this significant at the .05 level?

3. You think the average age for first marriage for a man is different from 27. In a random sample of 51 married men, the average age at the time of marriage was 28.1 with a standard deviation of 6.2. Is this finding statistically significant at the alpha = .02 level?

Hypothesis testing on two proportions

Step 1: State the hypotheses

Null: $\pi_1 - \pi_2 = 0$

Alt: $\pi_1 - \pi_2$ differs from 0

Step 2: Determine p_1, p_2 and p

p= (total successes)/(total sample size)

Check the assumptions

Step 3: Find the Z-score

$$Z = \frac{p_1 - p_2}{\sqrt{p(1-p)(\frac{1}{n_1} + \frac{1}{n_2})}}$$

Step 4: Determine the p-value

Step 5: Make the conclusion

Step 6: Make the concluding statement

Hypothesis testing on two proportions, in class example

You believe the proportion of boys who pass the math SOL will be higher than the girls. In two random samples of size 100 each, 88% of the boys passed, and 82% of the girls passed. Is this significant at the .05 level?

Define the variables/check the assumptions:

Hypotheses:

Determine the sample proportions (three of them):

Determine the test statistic:

P-value:

Conclusion:

Statement:

Determining the test statistic and p-value (for hypothesis testing on two proportions) on the TI 83/84

Press [STAT] and arrow over twice to TESTS
Select option 6: 2-PropZTest...
x1: State how many successes are in sample 1
n1: State the sample size of sample 1
x2: State how many successes in sample 2
n2: State the sample size of sample 2
p1: State whether the alternative is ≠, < or >
Select Calculate

Examples:
You believe the proportion of boys who pass the SOL test will be lower than girls in the third grade. A random sample of 50 boys and 64 girls was collected. 84% of the boys passed the SOL, whereas 85.9375% of the girls passed. Is this significant at the .05 level?

You have developed a new drug that you believe will reduce the incidence of strokes in high risk patients. A random sample of 35 people took the actual drug, and a random sample of 30 people took the placebo. 15 people on the actual drug had a stroke, and 14 people on the placebo had a stroke. Does this show, at significance level .01, that the drug is effective?

Hypothesis testing on two means with sigma known

State the hypotheses:

Null: $\mu_1 - \mu_2 = 0$

Alt: $\mu_1 - \mu_2$ differs from 0

Check the assumptions

Determine t:

$$Z = \frac{\overline{x}_1 - \overline{x}_2}{\sqrt{\left(\dfrac{\sigma_1^2}{n_1} + \dfrac{\sigma_2^2}{n_2}\right)}}$$

Find the p-value

Draw the conclusion

Make the concluding statement

Example:

Two machines are being compared to see if they are producing parts with different average sizes. Machine A has a known standard deviation of .04 inches, and machine B has a known standard deviation of .03 inches. 50 randomly selected parts from each machine were gathered, and their averages were reported. The parts from machine A had an average length of 10.02 inches, and the parts from machine B had an average of 9.99 inches. Is this difference significant at the .05 level?

Finding the test statistic and p-value (for hypothesis tests on two means with sigma known) on the TI 83/43

Press [STAT] and arrow over twice to TESTS
Select option 3: 2-SampZTest...
Select Stats
$\sigma 1$, $\sigma 2$: Enter the known standard deviations from the populations
$\bar{x}_1$, $\bar{x}_2$: Enter the sample means
n1, n2: Enter the sample sizes
Select whether the problems is $\neq$, < or >
Select Calculate

Examples:

SAT scores from School A are normally distributed with a known standard deviation of 302.4 points. SAT scores from School B are normally distributed with a known standard deviation of 298.5 points. You wish to see if there is a difference in the average scores of this year's students. A random sample of 15 students from school A had an average score of 1543.4, and a random sample of 10 students from school B had an average score of 1475.6. Is this significant at the .10 level?

Babies in the United States have normally distributed birth weights with a known standard deviation of 2.15 pounds. Babies in a third-world country have normally distributed birth weights with a known standard deviation of 2.59 pounds. You think the average birth weight is higher in the United States than it is in the third-world country. A random sample of 25 babies from the United States averaged at 7.23 pounds, and a random sample of 30 babies from the third-world country averaged at 6.97 pounds. Is this significant at the .05 level?

Hypothesis testing on two means with sigma unknown

State the hypotheses:

Null: $\mu_1 - \mu_2 = 0$

Alt: $\mu_1 - \mu_2$ differs from 0

Check the assumptions

Determine t:

$$t = \frac{\bar{X}_1 - \bar{X}_2}{\sqrt{(\frac{s_1^2}{n_1} + \frac{s_2^2}{n_2})}}$$

Determine the correct number of degrees of freedom:

Use the smaller of the two sample sizes to determine the df.

Find the range that contains the p-value.

Draw the conclusion

Make the concluding statement

In class example for hypothesis testing for two means with sigma unknown

You believe the students from school A have a higher average SAT score than students from school B. A random sample of 56 students from school A had an average score of 1697 with a standard deviation of 42. A random sample of 51 students from school B had an average of 1689 with a standard deviation of 39. Is this significant at the .05 level?

Define the variables/check the assumptions:

Hypotheses:

Test statistic:

P-value:

Conclusion:

Concluding statement:

Finding the test statistic and p-value (for hypothesis tests on two means with sigma known) on the TI 83/43

Press [STAT] and arrow over twice to TESTS
Select option 4: 2-SampTTest...
Select Stats
Sx1, Sx2: Enter the standard deviations from the samples
$\bar{x}1, \bar{x}2$: Enter the sample means
n1, n2: Enter the sample sizes
Select whether the problems is ≠, < or >
Pooled: No
Select Calculate

Examples:

The weight losses (in pounds) from randomly selected people from two different health clubs were recorded, and are as follows:

Club A:	12	8	14	18	22	4	9	14	16	17

Club B:	11	8	12	14	13	11	10	9

Does this show, at the .05 level, that the clubs have different average weight losses?

You believe the average GPA for first born children is higher than other members of the family. You randomly select 25 first born children and 38 non-first born children. The average GPA for first borns is 3.25 and for non-first borns is 3.16, with standard deviations of .13 and .08 respectively. Is this significant at the .01 level?

Mixed examples:

1. You believe the proportion of students getting an A in Professor Jones's class is higher than the proportion of students getting an A in Professor Smith's calss. In two random samples, 42 out of 60 students got an A in Professor Jones's class, and 25 out of 50 got an A in Professor Smith's class. Is this significant at the .05 level?

2. You believe the average IQ at School A is higher than the average IQ at School B. A random sample of 100 students at School A had an average IQ of 103.74 with a standard deviation of 15.13. A random sample of 78 students from School B had an average IQ of 99.85 with a standard deviation of 13.42. Is this significant at the .05 level?

3. You believe the average SOL score will be higher if students take practice SOL tests in advance. A random sample of 44 students took practice SOL tests and got an average score of 478.3. A random sample of 34 students didn't take practice tests and had an average score of 456.8. The known standard deviation for SOL tests is 39.4. Is this significant at the .01 level?

4. You believe the proportion of ADHD students receiving grades of C or lower will decrease with a certain curriculum. A random sample of 40 ADHD students took the new curriculum, and 57.5% received a C or lower. A random sample of 50 ADHD students took the regular curriculum, and 70% received a C or lower. Is this significant at the .05 level?

Mixed examples:

1. A study has shown that 18% of people speed when a police officer is in a patrol car at the side of the road. You believe the percentage will be higher if the patrol car is empty. A recent study had an empty police car parked at the side of the road. A random sample of 100 cars passing the empty patrol car showed that 21 of them were speeding. Does this show that a higher percentage of people will speed past an empty patrol car, at significance level .05?

2. The average age of CEO's of companies in America is 58.7. A random sample of 80 European CEO's showed an average age of 56.6, with a standard deviation of 6.8 years. Does this show, at significance level .01, that European CEO's are, on average, younger than American CEO's?

3. A machine is supposed to produce TV screens with an average diagonal length of 36 inches. The actual lengths of the screens produced are normally distributed with a known standard deviation of .03 inches. A random sample of 24 TV screens was taken, and their average was 36.018 inches. Does this show, at significance level .05, that the machine needs to be fixed?

4. A current study states that 4% of all college students admit to cheating on an exam. You believe the percentage is lower than that. A random sample of 125 people revealed that 5 have admitted to cheating on an exam. Does this show, at significance level .02, that fewer than 4% of all students have admitted to cheating?

5. SAT scores are normally distributed with a mean of 1500 and a known standard deviation of 300. You have created an SAT prep course that you think will increase the mean SAT score. You randomly select 30 high school students to take your course, and their average SAT score was 1575. Does this show, at significance level .05, that the average SAT score will be higher if all students completed this course?

6. A diet pill company claims that the average weight loss for a year will be 15 pounds on this product alone (diet and exercise not necessary). You think that sounds too good to be true and the actual average will be a lot lower. A random sample of 40 people took this pill for a year, and their average weight loss was 12.6 pounds with a standard deviation of 5.3 pounds. Does this provide evidence, at significance level .01, that the actual average is less than the claim? What about at significance level .05?

The relationship between confidence intervals and hypothesis tests

When creating a confidence interval, the range is centered about the point estimate. Sketch a 90% confidence interval on a normal curve:

90% of that interval is contained within the curve, which leaves 10% outside of it—5% above and 5% below.

If you conducted a one-sided hypothesis test, this would equate to an alpha of .05.

If you conducted a two-sided hypothesis test, this would equate to an alpha of .10

The rule:

When the confidence interval contains the hypothesized value, we have not shown a significant difference (fail to reject the null.) When the confidence interval does not contain the hypothesized value, we have shown a significant difference (reject the null.)

Examples:

1. You believe the mean IQ will increase after people complete an educational program. The current average is 100 with a known standard deviation of 15. In a random sample of 50 people, the average IQ was 103 after completing the program. Test this at the alpha = .05 level.

State the hypotheses. Notice the hypothesized value is 100, and this is a one-sided test.

If appropriate, conduct a hypothesis test.

Because this is a one sided test, use the table to determine the confidence level necessary to conduct the same test using an interval. Alpha = .05 equates to 90% confidence.

If appropriate, create a 90% confidence interval:

If the hypothesized value (100) is in the interval, that means we failed to show a difference (and we would fail to reject the null.) Do your two findings agree?

2. Left-handed and right-handed people were randomly selected to take a manual dexterity test to see if there was a difference in the average time needed to complete it. The sample of 46 right-handed people finished with an average time of 56.4 seconds with a standard deviation of 11.6 seconds. The sample of 22 left-handed people finished with an average time of 51.7 seconds with a standard deviation of 13.4 seconds. Use both a confidence interval and a hypothesis test to determine if this is significant at the alpha = .05 level.

3. You believe the proportion of students from VCU who pass the Praxis test is higher than a neighboring college. In a random sample of 200 VCU students who took the Praxis, 75% passed. At the other university, a random sample of 200 students showed that 64% passed. Use both a confidence interval and a hypothesis test to determine if this is significant at the alpha = .01 level.

Questions about the relationship between confidence intervals and hypothesis tests:

1. If a test is done and the hypotheses are:

$$H_0: \mu_1 - \mu_2 = 0$$
$$H_a: \mu_1 - \mu_2 > 0$$

And the p-value turns out to be .0632, does the 90% confidence interval contain 0?

2. If a test is done and the hypotheses are:

$$H_0: \mu = 14$$
$$H_a: \mu \neq 14$$

And the p-value turns out to be .0013, does the 95% confidence interval contain 14?

3. What confidence level would you need to create for your interval to equate to a two-sided hypothesis test at alpha = .01?

Binomial Distribution

Binomial distributions occur when there are two options for a situation, such as "pass/fail," or "true/false." For the following example, we will use "male/female."

One of those options needs to be deemed a success and the other a failure; for instance, we can declare being a female a success and being a male a failure in this example. Please note that this isn't a judgment call. In certain situations being diagnosed with cancer would be considered a success, depending on the wording of the question being asked.

For this particular question, you are asked: **A local university is 70% female. If you were to randomly select five students, what is the probability that all five will be female?** Since we are asking about females, we would call being female a success.

Using the binomial formula provided by the teacher, determine the probability that all five students would be female.

What is the probability that four out of the five would be female?

What is the probability that three out of the five would be female?

What is the probability that two out of the five would be female?

What is the probability that one out of the five would be female?

What is the probability that zero out of the five would be female?

How to determine binomial probabilities in the TI83:

To determine the probability of EXACTLY that amount:

Select [2nd][VARS] to get the DISTR function

Choose option 0: binompdf

Enter the following information: (sample size, probability of success, number of successes)

To determine the probability of THAT AMOUNT AND DOWN:

Select [2nd][VARS] to get the DISTR function

Choose option 0: binompdf

Enter the following information: (sample size, probability of success, number of successes)

Verify that the answers done by hand are correct.

Now, looking at the answers, determine the probability of selecting three or fewer females. You would add the probabilities of getting three, two, one or zero females. The computer can do this automatically if you change the cumulative from false to true.

Cumulative = false means the computer will calculate the probability of getting EXACTLY that number in the sample.

Cumulative = true means the computer will calculate the probability of getting THAT NUMBER AND DOWN.

It becomes difficult, then, to determine which number to plug in for number_s. You have to pay careful attention to the wording of the question. For instance, refer to the following questions:

Find the probability that three or fewer females would be in the sample.

Find the probability that fewer than three females would be in the sample.

Those are two very different questions. The first one asks for the probability of "three and down," whereas the second asks for the probability of "two and down."

The way to determine what number to plug in for number_s is to write down the number in question and the two surrounding numbers. Here, 3 is the number in question.

2   3   4

Then, point to each number and ask yourself if it qualifies. Let's use "three or fewer" females as our first example.

Does 2 qualify as three or fewer?   Yes.

Does 3 qualify as three or fewer?   Yes.

Does 4 qualify as three or fewer?   No.

So the transition takes place between 3 and 4. Draw a vertical line between those two numbers.

THE NUMBER THAT YOU TYPE IN FOR NUMBER_S IS ALWAYS, UNQUESTIONABLY, WITHOUT FAIL THE NUMBER TO THE LEFT OF YOUR CUTOFF.

So in this case, we would type in 3 as number_s. We would also type in true under cumulative to indicate to the computer that we'd like 'three and down.'

$$2 \quad 3 \quad 4$$

In the other instance, when you were asked about 'fewer than three,' you would apply the same procedure, only with a different result.

$$2 \quad 3 \quad 4$$

Does 2 qualify as fewer than three?   Yes.

Does 3 qualify as fewer than three?   No.

The transition comes between two and three, so draw a line there. In this case, you would type in 2 for number_s since it falls to the left of your cutoff.

What about the instance of "more than three?" Well, you would still draw the number in question and the two surrounding numbers.

$$2 \quad 3 \quad 4$$

Does 2 qualify as more than three? No

Does 3 qualify as more than three? No

Does 4 qualify as more than three? Yes

You will draw your line between 3 and 4, AND YOU WILL TYPE IN 3 AS NUMBER_S SINCE IT IS TO THE LEFT OF THAT CUTOFF. However, the computer is going to give you the probability for that cutoff and down. In order to find the probability above that cutoff, you would have to subtract the computer's answer from 1.

Binomial Activity

1. Suppose you have a trick deck of cards, which contains 65% black cards.

a) If you draw 10 cards from the deck, what is the probability that you will draw exactly 6 black cards?

b) What is the probability that 4 or fewer cards will be black if you draw 8 from the deck?

c) If you draw 15 cards, what is the probability that more than 11 will be black?

2. During the summer in Florida, it rains 78% of the days.

a) If you travel to Florida for 6 days, what is the probability it will rain less than 4 days?

b) If you stay for 12 days, what it the probability that it will rain 9 or more days?

c) If you spend 20 days there, what is the probability it will rain exactly 15 days?

Additional Binomial examples

1. An amusement park game gives you a prize if the ball lands on your birth month. You have an 8.33% chance of winning.

What is the probability that:

You will win exactly one game if you play 5 times?

You will win 5 times or fewer if you played 100 times?

You will not win if you play 3 times?

2. A certain high school claims that 90% of its graduates go on to college.

What is the probability that:

If you pick 15 students at random, more than 12 are going to college?

17 or fewer out of 20 will be going to college?

If you choose 50 students, 44 or more will go to college?

3. 26% of all children get tubes in their ears for ear infections.

What is the probability that:

Out of 65 children, 16 or more will have tubes?

Fewer than 10 out of 40 will have tubes?

Exactly 26 out of 100 will have tubes?

Chi Square

Chi-Square is used to compare categorical variables. The results of tests are often displayed in tables, and Chi-Square is used to determine if the differences between categories are statistically significant.

For instance, suppose a test was done to see if boys and girls are at differing reading levels in the second grade. You randomly select 40 children of each gender and determine if they passed or failed a grade-appropriate reading test. The results are as follows:

	Passed Test	Failed Test
Girls	25	15
Boys	19	21

Your null hypothesis in this study would be that boys and girls read at the same level. (Remember the premise in the null is always that things are equal.) The alternative would be that boys and girls read at different levels.

What you need to do first is to create what is called an "expected value" table. If the null is true—meaning that boys and girls read the same—and 44 children passed the test (as found in the above table), you would expect that 22 boys and 22 girls passed the test. If 36 students failed the test, you would expect that 18 of each gender failed.

PLEASE NOTE THAT THE REASON YOU EXPECT AN EQUAL NUMBER OF GIRLS AND BOYS TO PASS IS BECAUSE YOU HAD AN EQUAL NUMBER OF BOYS AND GIRLS IN THE SAMPLE. If we had twice as many girls as boys in the study, we would expect twice as many girls to pass as boys.

Anyway, our expected value table looks like this:

	Passed Test	Failed Test
Girls	22	18
Boys	22	18

How to determine Chi Square on the TI83

Create a matrix by selecting [2nd][x^{-1}]

Arrow over twice to EDIT

Select the desired matrix, typing in the dimensions. (# of rows X # of columns)

Type in the observed values (default is [A])

Repeat the process, typing in the expected values into a different matrix (default is [B])

Select [STAT] and arrow over twice to TESTS

Choose option C: X^2- Test

Type in the letter of the matrices (or rely on default)

Select Calculate

How to fill in the Chi Square expected value table

Determine the sums of the rows and columns:

	Passed Test	Failed Test	SUM
Girls	25	15	40
Boys	19	21	40
SUM	44	36	80

Remove the given data from the table, leaving the sums

	Passed Test	Failed Test	SUM
Girls			40
Boys			40
SUM	44	36	80

To fill in each cell, determine **(Row Total)(Column Total)/(Overall Total)**

	Passed Test	Failed Test	SUM
Girls			40
Boys			40
SUM	44	36	80

Chi-Square activity

1. The following table is the result of a study done to see if a relationship exists between participation in a reading program and the ability to read at grade level. A reading competency test was given to 56 students, some of whom had completed the program and some of whom had not.

	Passed Test	Failed Test
Participated in Program	19	7
Did Not Participate	14	16

State the hypotheses:

a) Fill in the expected value chart:

	Passed Test	Failed Test
Participated in Program		
Did Not Participate		

b) Determine the p-value

c) At significance level .05, does this provide evidence that a relationship exists between completing the program and reading at grade level?

2) A study was done to see the effectiveness of nicotine gum, the nicotine patch, and quitting cold turkey when trying to quit smoking. The following information was gathered.

	Quit	Did Not Quit
Gum	10	15
Patch	12	16
Cold Turkey	5	32

State the hypotheses;

a) Complete the expected value chart

	Quit	Did Not Quit
Gum		
Patch		
Cold Turkey		

b) Determine the p-value

c) At significance level .05, does this show evidence that there is a difference in the effectiveness of the methods?

Unit 1 practice examples

1. **Label the following as qualitative, discrete quantitative or continuous quantitative.**

A) The number of cars in the parking lot

B) Temperature

C) Gender

D) Average number of children in each family in the county

E) Student ID number

F) Number of classes you are taking this semester

2. **The Smithfield Elementary School PTA has 652 members and holds a meeting to discuss if funds should be spent on a renovation project. 438 members show up for the meeting. To get an idea of whether or not the members will elect to pass the renovation project during the vote next week, you stand outside the exit door of the meeting and ask every tenth person who comes out which way they plan to vote. In all you poll 43 members, 27 of whom plan to vote in favor of the renovation project. That equates to 62.79%.**

A) What is the population of interest?

B) What is the sampling frame?

C) What is the sample?

D) What method was used to collect the sample?

E) Label the following as parameters or statistics

    a. 652

    b. 43

    c. 27

    d. 62.79

3. **Which exists in the following scenarios: selection bias, response bias or non-response bias?**

A) 176 people agree to be in your study; 119 people return the survey

B) Your responses will be read by your boss

C) Your survey gets mailed to several U of R students when it is intended for VCU students only

D) The question reads, "Do you plan to vote for Bob Jones or do you plan to ruin America?

4. **State whether the numbers involved are nominal, ordinal or interval.**

A) Rate your hunger level on a scale of 1 to 10

B) The professor determines his grading scale

C) Contestants are numbered 1 – 20; list your three favorites

D) In a race, your bib number is 146

E) State which age range you fall into

F) Which number, 1 – 50, corresponds to the state you live in?

5. **Your goal is to select a sample of Walmart employees from your county to see how they feel about their working conditions. State the method used to collect the sample, and determine if the method is good or bad.**

A) You send out an email asking employees to respond.

B) You obtain a list of all employees and use every 25th name on the list.

C) You assign each employee with a code number and use software to randomly generate numbers.

D) You randomly pick two stores from the county and use every employee from those two stores.

E) You use every employee from every store.

F) You use all of the employees from the store closest to your house.

G) You randomly select five stores and use 10 randomly selected employees from each of those five stores.

H) You randomly select ten employees from every store in the county.

6. You want to test the effectiveness of a new pill designed to reduce the incidence of heart attacks in high-risk patients. To collect the sample of 100 patients, you randomly selected five states on which to focus. From within each of those five states, you randomly selected five counties (25). From within those counties you randomly selected two cardiologists (50), and from each cardiologist you randomly selected two patients, a male and a female (100).

Once the sample is collected, the group gets divided evenly into 50 men and 50 women. Within each gender group, half of the people are randomly selected to receive the pill; the remaining half receives the placebo. The percent of people who have heart attacks are recorded for each group, as is the severity of the heart attack on a scale of one to five.

A) Who is the population of interest?

B) What is the sampling frame?

C) Who are the experimental units?

D) What sample collection method was used?

E) What type of scale (nominal, ordinal, interval) was used to rate the heart attacks?

F) What are the factors?

G) How many levels are there?

H) Was this study single or double blind?

I) How many replications are in each group?

J) What is/are the treatment group/s?

K) What is/are the control group/s?

L) Is this completely randomized, block, or matched pairs design?

7. **State the type of design (block, completely randomized or matched pairs).**

A) In order to test the effectiveness of the nicotine patch, identical twins who smoke were randomly selected to receive one of two treatments. One of the twins would wear the real patch; the other twin would wear the placebo. At the end of the study, the percent of people who quit smoking in each group were compared.

B) In order to test the effectiveness of an after school program, two-hundred students from two local elementary schools were randomly selected (one-hundred from each.) Within each school, fifty students were randomly chosen to attend the after school program. The scores on a standardized test were then recorded and compared within each school.

C) A study was done to determine the effectiveness of an herbal supplement for the treatment of arthritis symptoms. A group of three-hundred people was randomly selected. Half that group was randomly chosen to take the real supplement; the rest got a placebo. At the end of three months, each group reported how well they felt and the results were compared.

8.

The weight losses, in pounds, for a group of 36 dieters were recorded and are as follows:

19	26	25	19	13	28	20	42	21	21	17	33
	23	23	24	26	26	18	38	18	12	27	15
	28	19	15	29	17	31	25	12	27	24	16
	30	24									

A) Create a stem and leaf plot for the data.

B) What shape is the data?

C) State the five number summary, range and IQR

D) Draw a box and whisker plot, being sure to denote the outliers.

9. The following scores are from a recent exam in a class with 20 students.

0	41	67	67	68	72	74	75	76	77	77
79	80	81	82	82	86	88	90	94		

A) **Without performing any calculations, predict how the mean will compare to the median.**

B) **What is the mean of the data set? The standard deviation?**

C) **What is the median?**

D) **Which, the mean or the median, is a better representation of center?**

E) **What is the 10% trimmed mean?**

F) **What variable represents the mean in this case? Is it a statistic or a parameter?**

10. Newborn baby weights are normally distributed with a mean of 7.3 pounds and a standard deviation of 2.1 pounds.

A) **Draw and label the normal curve for baby weights.**

B) 68% of all babies weight between...

10. Consider the following table:

Tablespoons of Fertilizer	Growth of Plant in cm
2	3.4
3	4.1
4	4.3
5	4.8
6	5.1
7	5.9
8	5.8

A) **Graph the scatterplot and describe the relationship.**

B) **Is "plant growth" a qualitative, discrete quantitative or continuous quantitative variable?**

C) **What is the correlation coefficient? Why does that value have the sign (positive or negative)?**

D) **What is the equation of the regression line?**

E) **What does the slope of that line mean?**

F) **What does the intercept of that line mean?**

G) **What can be the expected plant growth if 5.5 tablespoons of fertilizer are used?**

11. Consider the following table that relates hours spent studying to exam score:

Hours Studying	Grade
1.5	71
3	79
4.1	85
2.6	81
5.2	95
0	92

A) Create the scatterplot.
B) Is the point (0, 92) an outlier or an influential observation?

C) What is the correlation coefficient?

D) What is the coefficient of determination?

E) Would this graph be useful in making predictions?

F) What is the equation of the regression line?

G) What does the slope of that line mean?

H) What does the intercept of that line mean?

Now eliminate the point (0, 92) from the table.

I) Create the scatterplot
J) What is the correlation coefficient?

K) What is the coefficient of determination?

L) What is the equation of the regression line?

M) What does the slope of that line mean?

N) What does the intercept of that line mean?

12. Students in a large lecture class were asked three questions. Use the following information to determine the appropriate probabilities:

Question 1:   Have blue eyes: .3
              Have brown eyes: .5
Question 2:   Own exactly one dog: .4
              Own more than one dog: .12
Question 3:   Own a minivan: .2

What is the probability a person picked at random will:

a) have blue or brown eyes?

b) have blue eyes and one dog?

c) not own a minivan?

d) own at least one dog?

e) not own a dog?

f) brown eyes and a minivan?

Unit 2 Practice Examples

1. **Heights of five-year-old boys are distributed such that X~N(40,1.8). The units are inches.**

a) What percent of five year old boys are less than 3 feet (36 inches) tall?

b) Eddie is 41.5 inches tall. What percent of five-year-old boys are taller than him?

c) What height marks the cutoff for the top two percent?

d) The shortest 10% of five-year-old boys are below what height?

e) What percent of five-year-old boys are between 37 and 41 inches?

5. Race times, in minutes, are distributed such that X ~N(35.76, 6.53).

   a) What percent of runners finished with a time less than 40 minutes?

   b) Find two times such that the middle 60% of runners finished between those two times.

   c) Annie ran at the 42nd percentile. What was her time?

   d) Sue ran in 33.81 minutes; her sister ran in 36.25 minutes. What percent of runners fell between them?

   e) Find the times that mark the endpoints of the IQR.

5. Heights of males are distributed such that X~N(69.25, 2.93), where x is in inches.

a) What is the probability that a randomly selected male will be more than 72 inches tall?

b) What is the probability that a randomly selected group of 25 men will have an average height less than 68.5 inches tall? What enables you to answer that question?

c) What is the probability that a randomly selected group of 55 men will have an average height between 69 and 70 inches?

d) What is the probability that a randomly selected male will be less than 66 inches tall?

e) What does the sampling distribution look like for samples of size 72?

6. **47.6% of voters selected the democratic candidate in the most recent presidential election.**

a) What is the probability that a randomly selected group of 60 people will reveal that more than 50% of voters selected the democratic candidate?

b) What does the sampling distribution look like for random samples of size 100?

c) What is the probability that a randomly selected group of 10 people will have less than 45% voting democratic?

d) What is the probability that a randomly selected group of 500 people will have less than 50% democratic voters?

Unit 3 Practice Examples

1. A random sample of 41 statistics students spent an average of 83.7 minutes on statistics work every week. The standard deviation from the sample was 6.8 minutes. If appropriate, calculate and interpret a 95% confidence interval for the average weekly study time for all statistics students.

2. A random sample of ninety 8 year olds showed that 54.4% of them believe in Santa Claus. If appropriate, calculate and interpret a 90% confidence interval for the percent of all 8 year olds who believe in Santa.

3. A random sample of 25 professional baseball players had an average age of 26.2 with a standard deviation of 2.1 years. If appropriate, calculate and interpret a 99% confidence interval for the average age of all professional baseball players.

4. A researcher wants to determine if a people with a particular disease have a lower average IQ than the general population. IQ's are normally distributed with a known standard deviation of 15. A random sample of 20 people with the disease was selected, and their average IQ was 95. If appropriate, calculate and interpret a 90% confidence interval for the average IQ for all people with this disease.
   a) If the average IQ for the general public is 100, does this interval show the average IQ for people with this disease is lower than the general public?

5. 62 out of 80 randomly selected people do not answer the phone if their caller ID says "unavailable." If appropriate, calculate and interpret a 99% confidence interval for the proportion of all people who don't answer "unavailable" calls.

6. A machine produces TV screens, the lengths of which are normally distributed with a known standard deviation of .04 inches. A random sample of 12 TV screens was taken, and their lengths measured. The results were as follows.

27.02 26.96 27.08 27.1   27.05 27.06 27.03 27   27.04 27.1   26.99 27.01

If appropriate, calculate and interpret a 90% confidence interval for the average length of the TV screens. If they are supposed to have an average of 27 inches, does the machine need repair?

7. Two curricula are being compared to see which one yields higher average scores on the standardized tests at the end of the year. A random sample of 110 students who used curriculum A had an average SOL score of 476.5 with a sample standard deviation of 31.5. A random sample of 89 students from school B had an average score of 459.4 with a sample standard deviation of 38.6. Does this provide evidence, at 90% confidence, that the curricula have different averages?

8. The birth weights of babies born to women without diabetes are normally distributed with a known standard deviation of 2.09 pounds. The birth weights of babies born to women with diabetes have a known standard deviation of 1.2 pounds. A study was done to see if the average birth weights of these groups differ. A random sample of 35 babies born to women without diabetes was taken, and their average birth weight was 7.13 pounds. A random sample of 31 babies born to women with diabetes was taken, and their average birth weight was 9.6 pounds. Does this show, at 99% confidence, that the average birth weights are different?

9. A study was done to see if the proportion of education majors was different for two local state schools. School A's random sample of 190 students showed 16 people were education majors. At school B, a random sample of 215 students revealed that 22 were education majors. Does this provide evidence that the proportions are different for the two schools (at 90% confidence?)

10. A current drug is 80% effective, and you have a drug you believe is more effective. A random sample of 1000 people showed that 820 felt relief on your new drug. Is this significant at the .01 level?

11. You wish to see if adding an exercise program to a diet increases the average weight loss for participants. In two random samples of size 100 each, the average weight loss for exercisers was 28.4 pounds with a standard deviation of 9.2, and the average weight loss for non-exercisers was 25.7 with a standard deviation of 8.3. Is this significant at the .05 level?

12. You believe the average height for eight year olds is more than 54 inches. A random sample of 50 eight year olds had an average height of 55.2 inches with a standard deviation of 3.8 inches. Is this significant at the .05 level?

13. You believe the proportion of girls receiving advanced scores on this year's SOL is higher than the boys. In two random samples, 60 out of 400 girls tested at the advanced level, and 48 boys out of 400 tested at the advanced level. Is this significant at the .05 level?

14. A machine produces photo paper which is supposed to be, on average, 8.5 inches wide, but you are testing to see if the machine is now off. The machine has a known standard deviation of .06 inches. In a random sample of 64 sheets of paper, the average width was 8.52 inches. Is this significant at the .02 level? Does the machine need repair?

15. You have conducted a one-sided hypothesis test at the alpha = .05 level. You have found the test to be significant.

a) What level of confidence would you need for a CI to reflect the same findings?

b) Would the interval contain the hypothesized value?

c) Suppose the hypothesis test had been two sided...what percent confidence would you need?

16. A flipped coin will land on heads fifty percent of the time. If a coin is flipped 30 times, what is the probability that it will land on heads exactly 15 times?

17. The probability of any given flight leaving on time is .8. If you randomly select 15 flights, what is the probability that more than 11 of them will be on time?

18. A particular major consists of 30% females. If a random sample of 60 students is selected, what is the probability that less than 20 students will be female?

19. You wish to see if there is a difference in the SOL pass rates of students who use "common core" math and traditional math. The results are as follows:

	Passed	Did not pass
Common Core	57	12
Traditional	68	4

State the hypotheses and test to see if the difference is significant at the .05 level.

20. You wish to see if there is a difference in the effectiveness of a vaccine when given to males and females. The results are as follows:

	Effective	Ineffective
Males	47	22
Females	153	65

State the hypotheses and determine if the difference is significant at the .02 level.

21. The probability of contracting the flu after receiving a vaccine is .15. If a random sample of 200 vaccinated people was collected, what is the probability that 25 or fewer would contract the flu?

22. The probability of experiencing a side effect on a new drug is .08. If 1000 people are given the drug, what is the probability that more than 100 of them will experience a side effect?

23. The probability of a student requiring remediation is .06. If 200 students are randomly selected, what is the probability that exactly 11 of them will require remediation?

Answers to unit 1 practice examples

1.

A) Discrete quantitative

B) Continuous quantitative

C) Qualitative

D) Continuous quantitative

E) Qualitative

F) Discrete quantitative

2.

A) The entire PTA (All 652 members)

B) The 438 members who were at that meeting

C) The 43 members chosen

D) Systematic

E) a. Parameter

   b. Statistic

   c. Statistic

   d. Statistic

3.

A) Non-response bias

B) Response bias

C) Selection bias

D) Response bias

4.

A) Ordinal

B) Interval

C) Nominal

D) Nominal

E) Interval

F) Nominal

5.

A) Voluntary response; bad

B) Systematic; good

C) Simple Random; good

D) Cluster; bad

E) Census; good for accuracy, bad for practicality

F) Convenience; bad

G) Multistage; good for large populations

H) Stratified; good

6.

A) All people who are at high risk for a heart attack

B) People who see cardiologists

C) The 100 people in the study

D) Multistage

E) Ordinal

F) Gender, pill/placebo

G) 4

H) Double blind

I) 25

J) Males taking the real pill, females taking the real pill

K) Males taking the placebo, females taking the placebo

L) Block

7.

A) Matched pairs

B) Block

C) Completely randomized

8.

A)

```
1*  3,2,2

1·  9,9,7,8,8,5,9,5,7,6          1|9 = 19

2*  0,1,1,3,3,4,4,4

2·  6,5,8,6,6,7,8,9,5,7

3*  3,1,0

3·  8

4*  2
```

B)

(Slightly) skewed to the right

C) Min: 12

Q1: 18

Med: 23.5

Q3: 27

Max: 42

Range: 30

IQR: 9

H) Lower fence: $18 - 1.5(9) = 4.5$

Upper fence: $27 + 1.5(9) = 40.5$

Outlier = 42

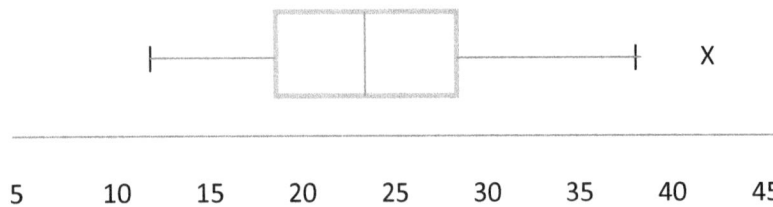

9.

A) The mean will be lower than the median

B) Mean = 72.8; std dev = 20.4311

C) Median = 77

D) Median is better because the mean is lowered by the outliers

E) 10% trimmed mean = 76.9375

F) μ The value is a parameter because it describes the population (the whole class)

10.

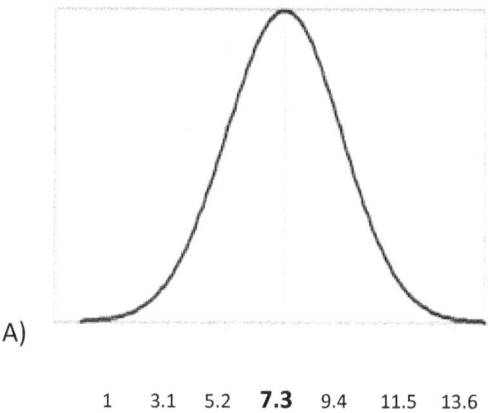

A)

1   3.1   5.2   **7.3**   9.4   11.5   13.6

B) 68% of babies weight between 5.2 and 9.4 pounds.

10.

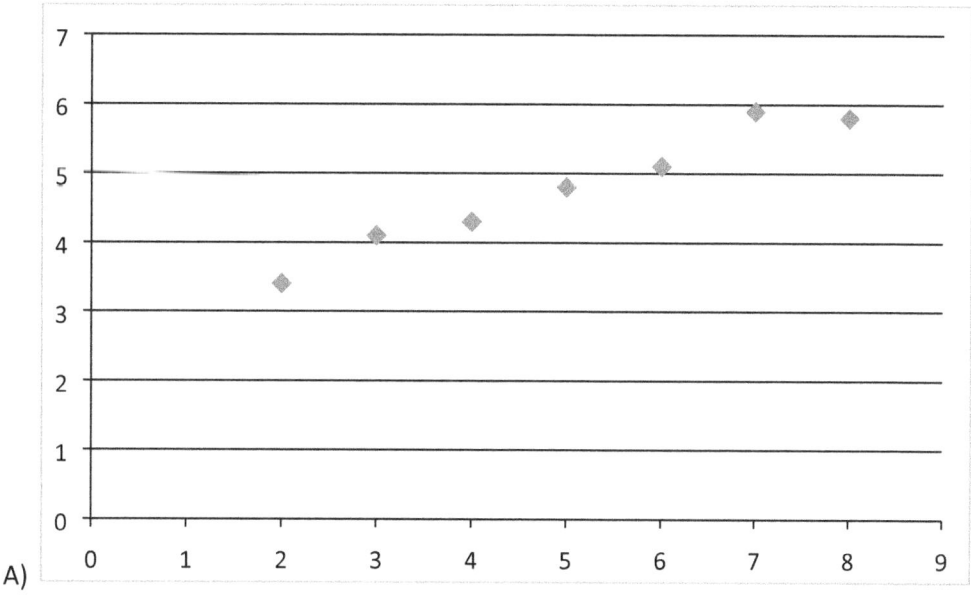

A)

There is a strong, positive linear relationship.

B) Continuous quantitative

C) r = .9809; as the amount of fertilizer increases, the amount of plant growth increases

D) y = 2.7 + .4143x

E) The slope means that for every tablespoon of fertilizer that gets added, the plant can be expected to be .4134 cm taller.

F) The intercept means that if no fertilizer were used, the plant can be expected to be 2.7 cm tall.

G) 4.9787 cm

11. A)

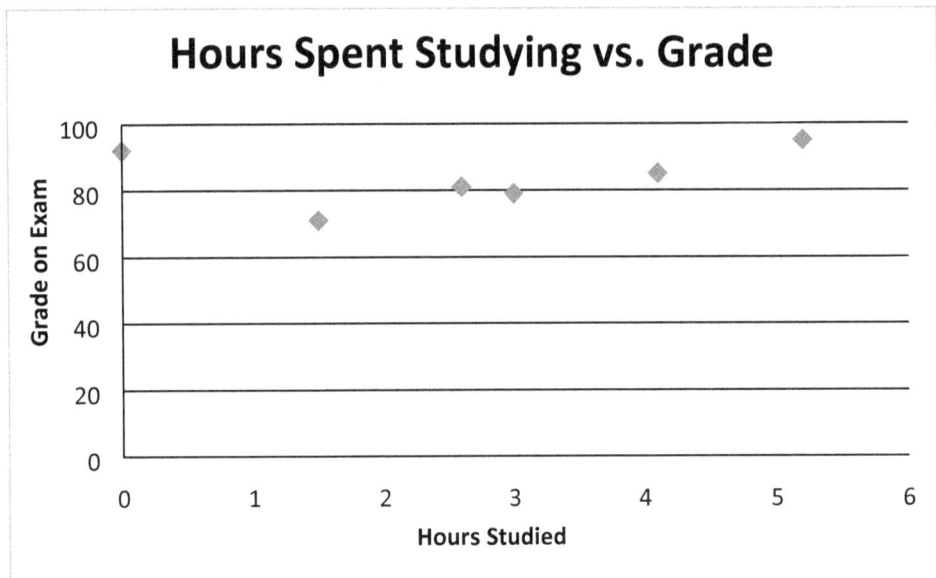

B) Influential observation

C) r = .2671

D) r^2 = .0713

E) Not useful

F) y = 80.3458 + 1.2759x

G) The slope of 1.2759 means that for every additional hour spent studying, you can expect your grade to increase by 1.2759 points.

H) The intercept of 80.3458 means if you do not study (x=0), you should get an 80.3458.

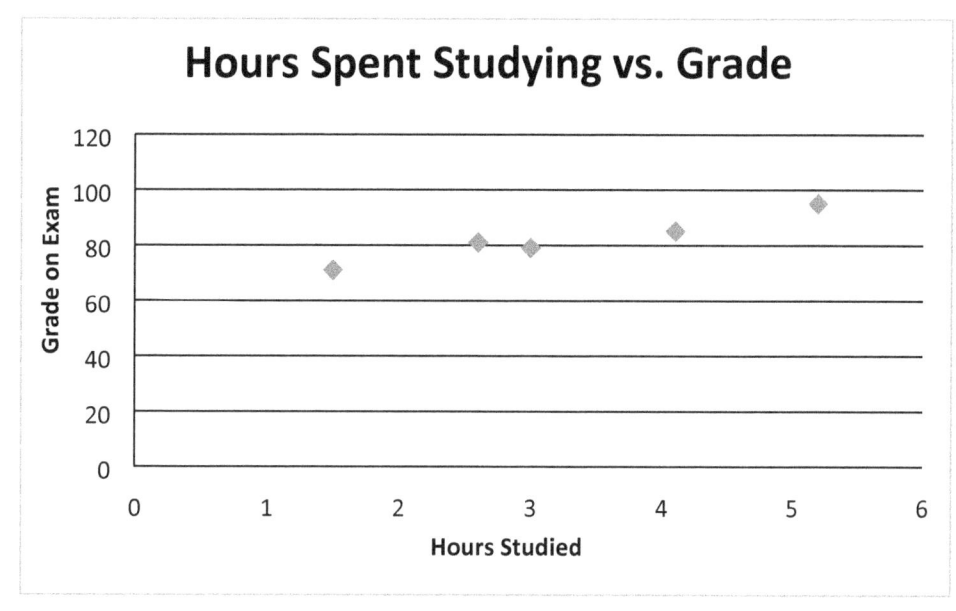

E)

J) .9721

K) .9449

L) y = 62.4745 + 6.0139x

M) For every additional hour spent studying, you can expect your exam score to increase by 6.0139 points.

N) The intercept means that if you do not study (x=0), you can expect to get a 62.4745

12.

a) .8

b) .12

c) .8

d) .52

e) .48

f) .1

Unit 2 Example Answers

1.

a) 1.3134%

b) 20.2328%

c) 43.6967 inches

d) 37.6932 inches

e) 66.2952%

2.

a) 74.1931%

b) 30.2642 – 41.2558 minutes

c) 34.4416 minutes

d) 14.7294%

e) 31.3556, 40.1644 minutes

3.

a) .1740

b) .0165; the sample was random, and the population was normally distributed.

c) .9712 - .2634 = .7078

d) .1337

e) X~N (69.25,0.3453)

7.

a) .3549

b) P~N(0.476, 0.0499)

c) Assumptions not met

d) .8587

Unit 3 Answers

1.

I am 95% confident that the average amount of time students spend studying statistics each week is between 81.554 and 85.846 minutes.

2.

I am 90% confident that between 45.81% and 63.079% of all eight-year-olds believe in Santa.

3.

I am 99% confident that the average age of professional baseball players is between 25.025 and 27.375 years.

4.

I am 90% confident that the average IQ for people with this disease is between 89.483 and 100.52. It does not appear the average is lower than the average for the general public.

5.

I am 99% confident that between 65.474% and 89.526% of all people do not answer the phone if the caller ID says "unavailable."

6.

I am 90% confident that the average width of the TV screens is between 27.018 and 27.056 inches. Yes, the machine producing them needs to be examined.

7.

Interval extends from (8.7053, 25.495)

I am 90% confident that the curricula produce different averages on the standardized tests at the end of the year.

8.

The interval extends from (-3.536, -1.404)

I am 99% confident that the average birth weight for babies born to women without diabetes is different from the average birth weight of babies born to women with diabetes.

9.

The interval extends from (-0.0656, 0.02936)

I am 90% confident that the proportion of education majors at university A is not different from the proportion of education majors at university B.

10.

H_0: $\pi = .8$
H_a: $\pi > .8$
Z = 1.5811
p-value = .0569
Fail to reject the null
There is insufficient evidence that the new drug is more effective.

11.

H_0: $\mu_1 - \mu_2 = 0$
H_a: $\mu_1 - \mu_2 > 0$
T = 2.1790
P-value = .0153
Reject the null
There is sufficient evidence that the average weight loss will increase with exercise.

12.

H_0: μ = 54
H_a: μ > 54
T = 2.2330
p-value = .0151
reject the null
There is sufficient evidence that the average height is more than 54 inches.

13.

H_0: $\pi_1 - \pi_2 = 0$
H_a: $\pi_1 - \pi_2 > 0$
Z = 1.2415
P-value = .1072
Fail to reject the null
There is insufficient evidence that the proportion of girls getting advanced scores is higher than boys

14.

H_0: μ = 8.5
H_a: μ ≠8.5
Z = 2.6667
p-value = .0077
reject the null
There is sufficient evidence that the average width is different from 8.5 inches.

15.
a) 90%
b) no
c) 95%

16. .1445

17. .6482

18. .6692

19. Null: Common core = traditional

Alt: Common core ≠ traditional

Expected values:

61.17	7.8298
63.83	8.1702

X^2 = 4.9064

P-value = .0268

Reject the null

There is sufficient evidence that a difference exists between common core and traditional curricula.

20.

Null: Male = female

Alt: Male ≠ female

Expected values:

48.084	20.916
151.92	66.084

X^2 = .1061

P-value = .7447

Fail to reject the null

There is insufficient evidence that a difference exists between males and females with respect to this vaccination.

21. .1876

22. .0101

23. .1173

How to perform tasks on the TI 83/84

Entering data:

Press [STAT] and then hit [ENTER] to select EDIT.

If any data exists in the column you are trying to use, you may arrow up to the title of the column (L_1 or L_2 for instance) and hit [CLEAR] then [ENTER].

Enter the values vertically down the columns, hitting [ENTER] between each value.

Making a histogram (univariate):

Once the data has been entered in one of the lists, hit [2ND] then [y=] to select the STATPLOT option.

Turn the Statplot ON.

Select the HISTOGRAM from the choices of TYPE.

Make sure you select the correct list (L_1 or L_2 for instance) for Xlist.

Adjust the WINDOW so that the graph will be visible on the screen.

Press [GRAPH]

Making a box and whisker plot (univariate):

Once the data has been entered in one of the lists, hit [2ND] then [y=] to select the STATPLOT option.

Turn the Statplot ON.

Select the BOX AND WHISKER from the choices of TYPE, designating whether you want outliers denoted or not.

Make sure you select the correct list (L_1 or L_2 for instance) for Xlist.

Choose which mark you would like if you want outliers denoted.

Adjust the WINDOW so that the graph will be visible on the screen.

Press [GRAPH]

Making a scatterplot or line graph (bivariate):

Once the data has been entered in two of the lists, hit [2ND] then [y=] to select the STATPLOT option.

Turn the Statplot ON.

Select the LINE GRAPH or SCATTERPLOT option from the choices of TYPE.

Make sure you select the correct list (L_1 or L_2 for instance) for Xlist.

Make sure you select the correct list (L_1 or L_2 for instance) for Ylist

Choose which mark you would like.

Adjust the WINDOW so that the graph will be visible on the screen.

Press [GRAPH]

Obtaining descriptive statistics (univariate):

Enter data into one of the lists.

Press the [STAT] button again and arrow over to CALC.

Select option 1 and hit [ENTER].

Select which list (L_1 or L_2 for instance) you would like analyzed. L_1 is the default if left blank.

Hit [ENTER]

You will be given the mean, standard deviation and five number summary.

Determining correlation coefficient, coefficient of determination and regression equation (bivariate):

Enter the data in two of the lists.

Select [2ND] and [0] to access the CATALOG.

Select DiagnosticOn and hit [ENTER] twice. (The word DONE should appear).

Press the [STAT] button again and arrow over to CALC.

Select option 8 and hit [ENTER].

Select which lists (L_1 or L_2 for instance) you would like analyzed, separated by a comma. The first list entered is the X (independent) variable. $L_1 = x$ and $L_2 = y$ are the defaults if left blank.

Hit [ENTER].

You will see r, r^2, and the slope and the intercept of the regression equation.

How to determine expected value and standard deviation on the TI 83/84

Press [STAT] and [ENTER] to edit the lists.

In L_1 type in the possible values of x

In L_2 type in the corresponding probabilities. If you type in those values as fractions, the calculator will automatically convert them to decimals.

Press [STAT] and arrow over to CALC.

Select option 1 for 1-Var Stats, typing in L_1, L_2 after.

Hit [ENTER]

$\bar{x}$ will tell you the expected value

σx will tell you the standard deviation.

To determine probabilities of obtaining certain values given a population mean and standard deviation:

For individuals:

Hit [STAT] and arrow over twice to TESTS.

Select option 1 and hit [ENTER]

Input: Stats

μ_0 = population mean

σ = population standard deviation

$\bar{x}$ = the individual value

n = 1

Determine if you want $\neq$, < or >

Select calculate

For sample means:

Hit [STAT] and arrow over twice to TESTS.

Select option 1 and hit [ENTER]

Input: Stats

μ_0= population mean

σ = population standard deviation

$\bar{x}$ = the sample mean

n = the sample size

Determine if you want $\neq$, $<$ or $>$

Select calculate

Finding a confidence interval for a single proportion on the TI 83/84

Press [STAT] and arrow over twice for TESTS.
Select A: 1-PropZInt
x: number of successes
n: sample size
C-Level: desired confidence level
Select Calculate

Finding confidence intervals on single means with sigma known on the TI 83/84

Press [STAT] and [ENTER] to put the information into a list
Press [STAT] and arrow over twice to TESTS
Choose option 7: ZInterval
Select Data
σ: Enter standard deviation
List: Choose which list contains the data
Freq: 1
C-level: state the confidence level
Select Calculate

Finding confidence intervals on single means with sigma unknown on the TI 83/84

Press [STAT] and arrow over twice to TESTS
Choose option 8: TInterval
Select Stats to input the mean and standard error
$\bar{x}$: sample mean
sx: standard error
n: sample size
C-level: state the confidence level
Select Calculate

Computing confidence intervals on the difference of two proportions on the TI 83/84:

Select [STAT] and arrow over twice to TESTS.

Choose option B: 2-PropZInt…

x1: The number of successes in sample 1
n1: The sample size of sample 1
x2: The number of successes in sample 2
n2: The sample size of sample 2

Select the confidence level in C-Level
Select Calculate

Determining a confidence level for a difference of two means (with sigma known) on the TI 83/84

Press [STAT] and arrow over twice to TESTS
Choose option 9: 2-SampZInt
Choose Stats
$\sigma 1$: Enter the known standard deviation for sample 1
$\sigma 2$: Enter the known standard deviation for sample 2
$\bar{x}1$: Enter the sample average for sample 1
n1: Enter the sample size for sample 1
$\bar{x}2$: Enter the sample average for sample 2
n2: Enter the sample size for sample 2
C-Level: Enter the desired confidence level
Select calculate.

Determining a confidence level for a difference of two means (with sigma unknown) on the TI 83/84

Press [STAT] and arrow over twice to TESTS
Choose option 0: 2-SampTInt
Choose Stats
$\bar{x}1$: Enter the sample average for sample 1
$sx1$: Enter the standard deviation from sample 1
n1: Enter the sample size for sample 1
$\bar{x}2$: Enter the sample average for sample 2
$sx2$: Enter the standard deviation from sample 2
n2: Enter the sample size for sample 2
C-Level: Enter the desired confidence level
Pooled: select No
Select calculate.

Finding the test statistic and p-value (for hypothesis testing on proportions) on the TI 83/84

Press [STAT] and arrow over twice to TESTS
Select option 5: 1-PropZTest
P0: enter the value for π
X: enter the number of successes in the sample
N: enter the sample size
State whether the alternative was $\neq$, $>$ or $<$
Select Calculate

How to obtain the test statistic and p-value (for hypothesis testing on a single with sigma known) on the TI 83/84

Press [STAT] and then arrow over twice for TESTS
Choose option 1: Z-Test…
Select Stats
$\mu 0$: Population mean from the null
σ: Population standard deviation
$\bar{x}$: Sample mean
n: Sample size
State whether the alternative is a $\neq$, $<$ or $>$
Select Calculate

How to obtain the test statistic and p-value (for hypothesis testing on a single with sigma unknown) on the TI 83/84

Press [STAT] and then arrow over twice for TESTS
Choose option 2: T-Test…
Select Stats
$\mu 0$: Population mean from the null
$\bar{x}$: Sample mean
Sx: Standard deviation from the sample
n: Sample size
State whether the alternative is a $\neq$, < or >
Select Calculate

Determining the test statistic and p-value (for hypothesis testing on two proportions) on the TI 83/84

Press [STAT] and arrow over twice to TESTS
Select option 6: 2-PropZTest…
x1: State how many successes are in sample 1
n1: State the sample size of sample 1
x2: State how many successes in sample 2
n2: State the sample size of sample 2
p1: State whether the alternative is $\neq$, < or >
Select Calculate

Finding the test statistic and p-value (for hypothesis tests on two means with sigma known) on the TI 83/43

Press [STAT] and arrow over twice to TESTS
Select option 3: 2-SampZTest…
Select Stats
$\sigma 1$, $\sigma 2$: Enter the known standard deviations from the populations
$\bar{x}_1$, $\bar{x}_2$: Enter the sample means
n1, n2: Enter the sample sizes
Select whether the problems is $\neq$, < or >
Select Calculate

Finding the test statistic and p-value (for hypothesis tests on two means with sigma known) on the TI 83/43

Press [STAT] and arrow over twice to TESTS
Select option 4: 2-SampTTest...
Select Stats
Sx1, Sx2: Enter the standard deviations from the samples
$\bar{x}_1, \bar{x}_2$: Enter the sample means
n1, n2: Enter the sample sizes
Select whether the problems is ≠, < or >
Pooled: No
Select Calculate

How to determine binomial probabilities in the TI83:

To determine the probability of EXACTLY that amount:

Select [2nd][VARS] to get the DISTR function

Choose option 0: binompdf

Enter the following information: (sample size, probability of success, number of successes)

To determine the probability of THAT AMOUNT AND DOWN:

Select [2nd][VARS] to get the DISTR function

Choose option 0: binompdf

Enter the following information: (sample size, probability of success, number of successes)

How to determine Chi Square on the TI83

Create a matrix by selecting [2nd][x^{-1}]

Arrow over twice to EDIT

Select the desired matrix, typing in the dimensions. (# of rows X # of columns)

Type in the observed values (default is [A])

Repeat the process, typing in the expected values into a different matrix (default is [B])

Select [STAT] and arrow over twice to TESTS

Choose option C: X^2- Test

Type in the letter of the matrices (or rely on default)

Select Calculate

t*

df	.25	.20	.15	.10	.05	.025	.02	.01	.005	.0025	.001	.0005
1	1.000	1.376	1.963	3.078	6.314	12.71	15.89	31.82	63.66	127.3	318.3	636.6
2	0.816	1.061	1.386	1.886	2.920	4.303	4.849	6.965	9.925	14.09	22.33	31.60
3	0.765	0.978	1.250	1.638	2.353	3.182	3.482	4.541	5.841	7.453	10.21	12.92
4	0.741	0.941	1.190	1.533	2.132	2.776	2.999	3.747	4.604	5.598	7.173	8.610
5	0.727	0.920	1.156	1.476	2.015	2.571	2.757	3.365	4.032	4.773	5.893	6.869
6	0.718	0.906	1.134	1.440	1.943	2.447	2.612	3.143	3.707	4.317	5.208	5.959
7	0.711	0.896	1.119	1.415	1.895	2.365	2.517	2.998	3.499	4.029	4.785	5.408
8	0.706	0.889	1.108	1.397	1.860	2.306	2.449	2.896	3.355	3.833	4.501	5.041
9	0.703	0.883	1.100	1.383	1.833	2.262	2.398	2.821	3.250	3.690	4.297	4.781
10	0.700	0.879	1.093	1.372	1.812	2.228	2.359	2.764	3.169	3.581	4.144	4.587
11	0.697	0.876	1.088	1.363	1.796	2.201	2.328	2.718	3.106	3.497	4.025	4.437
12	0.695	0.873	1.083	1.356	1.782	2.179	2.303	2.681	3.055	3.428	3.930	4.318
13	0.694	0.870	1.079	1.350	1.771	2.160	2.282	2.650	3.012	3.372	3.852	4.221
14	0.692	0.868	1.076	1.345	1.761	2.145	2.264	2.624	2.977	3.326	3.787	4.140
15	0.691	0.866	1.074	1.341	1.753	2.131	2.249	2.602	2.947	3.286	3.733	4.073
16	0.690	0.865	1.071	1.337	1.746	2.120	2.235	2.583	2.921	3.252	3.686	4.015
17	.0689	0.863	1.069	1.333	1.740	2.110	2.224	2.567	2.898	3.222	3.646	3.965
18	0.688	0.862	1.067	1.330	1.734	2.101	2.214	2.552	2.878	3.197	3.611	3.922
19	0.688	.0861	1.066	1.328	1.729	2.093	2.205	2.539	2.861	3.174	3.579	3.883
20	0.687	0.860	1.064	1.325	1.725	2.086	2.197	2.528	2.845	3.153	3.552	3.850
21	0.686	0.859	1.063	1.323	1.721	2.080	2.189	2.518	2.831	3.135	3.527	3.819
22	0.686	0.858	1.061	1.321	1.717	2.074	2.183	2.508	2.819	3.119	3.505	3.792
23	0.685	0.858	1.060	1.319	1.714	2.069	2.177	2.500	2.807	3.104	3.485	3.768
24	0.685	0.857	1.059	1.318	1.711	2.064	2.172	2.492	2.797	3.091	3.467	3.745
25	0.684	0.856	1.058	1.316	1.708	2.060	2.176	2.485	2.787	3.078	3.450	3.725
26	0.684	0.856	1.058	1.315	1.706	2.056	2.162	2.479	20779	3.067	3.435	3.707
27	0.684	0.855	1.057	1.314	1.703	2.052	2.158	2.743	2.771	3.057	3.421	3.690
28	0.683	.0855	1.056	1.313	1.701	2.048	2.154	2.467	2.763	3.047	3.408	3.674
29	0.683	0.854	1.055	1.311	1.699	2.045	2.150	2.462	2.756	3.038	3.396	3.659
30	0.683	0.854	1.055	1.310	1.697	2.042	2.147	2.457	2.750	2.971	3.307	3.551
40	0.681	0.851	1.050	1.303	1.684	2.021	2.123	2.423	2.704	2.971	3.307	3.551
50	0.679	0.849	1.047	1.299	1.676	2.009	2.109	2.403	2.678	2.937	3.261	3.496
60	0.679	0.848	1.045	1.296	1.671	2.000	2.099	2.390	2.660	2.915	3.232	3.460
80	0.678	0.846	1.043	1.292	1.664	1.990	2.088	2.374	2.639	2.887	3.195	3.416
100	0.677	0.845	1.042	1.290	1.660	1.984	2.081	2.364	2.626	2.871	3.174	3.390
1000	0.675	0.842	1.037	1.282	1.646	1.962	2.056	2.330	2.581	2.813	3.098	3.300
z*	0.674	0.841	1.036	1.282	1.645	1.960	2.054	2.326	2.576	2.807	3.091	3.291
	50%	60%	70%	80%	90%	95%	96%	98%	99%	99.5%	99.8%	99.9%

Confidence Level

Z table

z	.00	.01	.02	.03	.04	.05	.06	.07	.08	.09
-3.4	.0003	.0003	.0003	.0003	.0003	.0003	.0003	.0003	.0003	.0002
-3.3	.0005	.0005	.0005	.0004	.0004	.0004	.0004	.0004	.0004	.0003
-3.2	.0007	.0007	.0006	.0006	.0006	.0006	.0006	.0005	.0005	.0005
-3.1	.0010	.0009	.0009	.0009	.0008	.0008	.0008	.0008	.0007	.0007
-3.0	.0013	.0013	.0013	.0012	.0012	.0011	.0011	.0011	.0010	.0010
-2.9	.0019	.0018	.0018	.0017	.0016	.0016	.0015	.0015	.0014	.0014
-2.8	.0026	.0025	.0024	.0023	.0023	.0022	.0021	.0021	.0020	.0019
-2.7	.0035	.0034	.0033	.0032	.0031	.0030	.0029	.0028	.0027	.0026
-2.6	.0047	.0045	.0044	.0043	.0041	.0040	.0039	.0038	.0037	.0036
-2.5	.0062	.0060	.0059	.0057	.0055	.0054	.0052	.0051	.0049	.0048
-2.4	.0082	.0080	.0078	.0075	.0073	.0071	.0069	.0068	.0066	.0064
-2.3	.0107	.0104	.0102	.0099	.0096	.0094	.0091	.0089	.0087	.0084
-2.2	.0139	.0136	.0132	.0129	.0125	.0122	.0119	.0116	.0113	.0110
-2.1	.0179	.0174	.0170	.0166	.0162	.0158	.0154	.0150	.0146	.0143
-2.0	.0228	.0222	.0217	.0212	.0207	.0202	.0197	.0192	.0188	.0183
-1.9	.0287	.0281	.0274	.0268	.0262	.0256	.0250	.0244	.0239	.0233
-1.8	.0359	.0351	.0344	.0336	.0329	.0322	.0314	.0307	.0301	.0294
-1.7	.0446	.0436	.0427	.0418	.0409	.0401	.0392	.0384	.0375	.0367
-1.6	.0548	.0537	.0526	.0516	.0505	.0495	.0485	.0475	.0465	.0455
-1.5	.0668	.0655	.0643	.0630	.0618	.0606	.0594	.0582	.0571	.0559
-1.4	.0808	.0793	.0778	.0764	.0749	.0735	.0721	.0708	.0694	.0681
-1.3	.0968	.0951	.0934	.0918	.0901	.0885	.0869	.0853	.0838	.0823
-1.2	.1151	.1131	.1112	.1093	.1075	.1056	.1038	.1020	.1003	.0985
-1.1	.1357	.1335	.1314	.1292	.1271	.1251	.1230	.1210	.1190	.1170
-1.0	.1587	.1562	.1539	.1515	.1492	.1469	.1446	.1423	.1401	.1379
-0.9	.1841	.1814	.1788	.1762	.1736	.1711	.1685	.1660	.1635	.1611
-0.8	.2119	.2090	.2061	.2033	.2005	.1977	.1949	.1922	.1984	.1867
-0.7	.2420	.2389	.2358	.2327	.2296	.2266	.2236	.2206	.2177	.2148
-0.6	.2743	.2709	.2676	.2643	.2611	.2578	.2546	.2514	.2483	.2451
-0.5	.3085	.3050	.3015	.2981	.2946	.2912	.2877	.2843	.2810	.2776
-0.4	.3446	.3409	.3372	.3336	.3300	.3264	.3228	.3192	.3156	.3121
-0.3	.3821	.3783	.3745	.3707	.3669	.3632	.3594	.3557	.3520	.3483
-0.2	.4207	.4168	.4129	.4090	.4052	.4013	.3974	.3936	.3897	.3859
-0.1	.4602	.4562	.4522	.4483	.4443	.4404	.4364	.4325	.4286	.4247
-0.0	.5000	.4960	.4920	.4880	.4840	.4801	.4761	.4721	.4681	.4641

z	.00	.01	.02	.03	.04	.05	.06	.07	.08	.09
0.0	.5000	.5040	.5080	.5120	.5160	.5199	.5329	.5279	.5319	.5359
0.1	.5398	.5438	.5478	.5517	.5557	.5596	.5636	.5675	.5714	.5753
0.2	.5793	.5832	.5871	.5910	.5948	.5987	.6026	.6064	.6103	.6141
0.3	.6179	.6217	.6255	.6293	.6331	.6368	.6406	.6443	.6480	.6517
0.4	.6554	.6591	.6628	.6664	.6700	.6736	.6772	.6808	.6844	.6879
0.5	.6915	.6950	.6985	.7019	.7054	.7088	.7123	.7157	.7190	.7224
0.6	.7257	.7291	.7324	.7357	.7389	.7422	.7454	.7486	.7517	.7549
0.7	.7580	.7611	.7642	.7673	.7704	.7734	.7764	.7794	.7823	.7852
0.8	.7881	.7910	.7939	.7967	.7995	.8023	.8051	8078	.8106	.8133
0.9	.8159	.8186	.8212	.8238	.8264	.8289	.8315	.8340	.8365	.8389
1.0	.8413	.8438	.8461	.8485	.8508	.8531	.8554	.8577	.8599	.8621
1.1	.8643	.8665	.8686	.8708	.8729	.8749	.8770	.8790	.8810	.8830
1.2	.8849	.8869	.8888	.8907	.8925	.8944	.8962	.8980	.8997	.9015
1.3	.9032	.9049	.9066	.9082	.9099	.9115	.9131	.9147	.9162	.9177
1.4	.9192	.9207	.9222	.9236	.9251	.9265	.9279	.9292	.9306	.9319
1.5	.9332	.9345	.9357	.9370	.9382	.9394	.9406	.9418	.9429	.9441
1.6	.9452	.9463	.9474	.9484	.9495	.9505	.9515	.9525	.9535	.9545
1.7	.9554	.9564	.9573	.9582	.9591	.9599	.9608	.9616	.9625	.9633
1.8	.9641	.9649	.9656	.9664	.9671	.9678	.9686	.9693	.9699	.9706
1.9	.9713	.9719	.9726	.9732	.9738	.9744	.9750	.9756	.9761	.9767
2.0	.9772	.9778	.9783	.9788	.9793	.9798	.9803	.9808	.9812	.9817
2.1	.9821	.9826	.9830	.9834	.9838	.9842	.9846	.9850	.9854	.9857
2.2	.9861	.9864	.9868	.9871	.9875	.9878	.9881	.9884	.9887	.9890
2.3	.9893	.9896	.9898	.9901	.9904	.9906	.9909	.9911	.9913	.9916
2.4	.9918	.9920	.9922	.9925	.9927	.9929	.9931	.9932	.9934	.9936
2.5	.9938	.9940	.9941	.9943	.9945	.9946	.9948	.9949	.9951	.9952
2.6	.9953	.9955	.9956	.9957	.9959	.9960	.9961	.9962	.9963	.9964
2.7	.9965	.9966	.9967	.9968	.9969	.9970	.9971	.9972	.9973	.9974
2.8	.9974	.9975	.9976	.9977	.9977	.9978	.9979	.9979	.9980	.9981
2.9	.9981	.9982	.9982	.9983	.9984	.9984	.9985	.9985	.9986	.9986
3.0	.9987	.9987	.9987	.9988	.9988	.9989	.9989	.9989	.9990	.9990
3.1	.9990	.9991	.9991	.9991	.9992	.9992	.9992	.9992	.9993	.9993
3.2	.9993	.9993	.9994	..9994	.9994	.9994	.9994	.9995	.9995	.9995
3.3	.9995	.9995	.9995	.9996	.9996	.9996	.9996	.9996	.9996	.9997
3.4	.9997	.9997	.9997	.9997	.9997	.9997	.9997	.9997	.9997	.9998